Oligopoly

Tönu Puu

Oligopoly

Old Ends - New Means

Springer

Prof. Dr. Tönu Puu
Umeå University
CERUM
90187 Umeå
Sweden
tonu.puu@econ.umu.se

ISBN 978-3-642-15963-3 e-ISBN 978-3-642-15964-0
DOI 10.1007/978-3-642-15964-0
Springer Heidelberg Dordrecht London New York

Cover design: WMXDesign

Printed on acid-free paper

Springer is part of Springer Science+Business Media (www.springer.com)

Preface

The topic for this volume is classical oligopoly theory as it took shape around 1940. In the 100 years after Augustin Cournot first formulated the oligopoly problem in 1838, many intriguing developments of the theory were suggested. For a long period after 1940 many of these were left as loose ends. Oligopoly was a candidate for the application of game theory, so it came under the spell of the probabilistic optimum and equilibrium settings of the latter.

As a consequence, the deterministic dynamic aspects disappeared from the agenda until 1976, when David Rand suggested that oligopoly could be a candidate from economics for the application of the new dynamics based on topology and numerical methods that had been growing for quite a while, and had given exciting new perspectives on issues such morphogenesis in plants, the stability of the solar system, or prediction in meteorology.

After this beginning, the study of nonlinear dynamic systems in economics has been a fast-growing field, and oligopoly theory has been a substantial part of it, exploring many new and varied ideas. However, the researchers, often of an interdisciplinary origin, have been more interested in exploring analogies to models from other sciences known to produce spectacular results, than in exploring the actually existent heritage on the topic from economics proper. Main stream economists offered no competition as their interest was in perfect markets, optimal structures, equilibrium and stability, rather than in imperfections, multistability, and dynamics.

As a consequence the rich heritage from around 1940 was left as loose ends, so the present author feels it is worthwhile to reconsider and concentrate just on these loose ends in the light of the new dynamics.

Names that recur frequently in the sequel are von Stackelberg, Robinson, Palander, Hotelling, Wald, Lerner and Singer, in contributions that are sadly neglected.

The present book draws on numerous papers by its author with collaborators with provenance from the field of mathematics, whose work has been indispensable for the results. Even if the present author alone is responsible for the formulation of the models, the interaction with professional mathematicians was indispensable for finding out which problems are possible to analyse and which are just too complicated; and which problems are likely to produce interesting results and which just result in trivialities. Such interaction back and forth, helps the economist to develop the model formulation.

The author wants to thank several of these collaborators, above all, Prof. *Laura Gardini*, Urbino, Profs. *Manuel Rùiz* and *José Cànovas*, Cartagena, Drs. *Iryna Sushko* and *Anastasiia Panchuk*, Kiev, Prof. *Anna Agliari*, Milano, and Dr. *Fabio Tramontana*, Ancona. Given the published articles focus on mathematics, with sometimes much too little economics intuition to explain the models, the present volume concentrates on economics, and aims at the traditional economist style of presentation, including interpretations of intermediate results whenever this is possible, rather than the Lemma, Theorem, Q.E.D. style.

Umeå, Sweden
March 2010 Tönu Puu

Contents

1 Introduction

Early History of Mathematical Economics

According to Joseph Schumpeter's "*History of Economic Analysis*" from 1954 - the still unsurpassed source book for economics, mathematical modelling in the field had two roots. The first was Johann Heinrich von Thünen's theory of land use dating from 1827, implicitly containing the most general theory of the emergence of specialization and trade ever presented up to now, as it was based on the mathematical principle of transversality alone; much more general than the Ricardian comparative advantages or the later Heckscher–Ohlin theory which also was based on immobile resources. As a historical irony the theory has not to this day been recognized as anything more than a theory in agricultural economics.

The second attempt, on oligopoly by Augustin Cournot from 1838, has, however kept theoretical economists busy to this day. Later in the century, more precisely in 1883, Bertrand made an attempt at killing that other embryo of mathematical modelling. Bertrand's real objective was to deny the possibility of mathematical modelling in a field such as economics, as he also attacked Léon Wálras's much later theory of general equilibrium in the same article.

However, Bertrand launched some substantial objections to Cournot's theory of competition among "the few", as Fellner (1949) later very descriptively called it. The argument was quite confused, so it seems to be an undeserved honour to call the alternative oligopoly theory "Bertrand oligopoly" as if it were at all comparable to Cournot's penetrating analysis.. Schumpeter expresses his "*doubt whether*" Bertrand "*would have made much impression if .... others had not .... repudiated Cournot's solution*".

But, now we run ahead of history without even having stated the arguments yet.

T. Puu, *Oligopoly: Old Ends – New Means*, DOI 10.1007/978-3-642-15964-0_1,

Cournot Oligopoly

As early as in 1838 Cournot, a pure mathematician, presented all the elements of oligopoly theory. Oligopoly, competition among few sellers, is placed between monopoly, one seller, and perfect competition. From microeconomics textbooks, in perfect competition there are so many, implicitly taken as small size firms, that, even if market price depends on total supply, each firm takes price as virtually fixed, because within its possible range of supply variation, it cannot perceptibly influence either total supply or market price. Hence the firm in perfect competition applies marginal cost pricing. If the firms have different cost functions, those with lower costs get a positive profit, but the firm on the margin earns nothing; when the firms are identical and there is free entry for them, then this holds for all firms.

On the opposite end of the spectrum is monopoly; there is one single supplier who takes explicit account of the market demand function, and deliberately limits its supply in order to charge a higher monopoly price, at the disadvantage of the consumers. Marginal costs are equated, not to price, but to marginal revenue, which as a rule is lower.

Oligopoly and duopoly, the case of two competitors, is the intermediate case in terms of substance, but not in terms of analysis. It is more complicated than either monopoly or competition. The oligopolist, like the monopolist, takes account of market demand and attempts to equate marginal revenue to marginal cost. But now the marginal revenue for each firm depends, not only on its own supply, but on the supplies of all the competitors as well, because each of them is large enough to influence market price. Therefore, to calculate an optimal decision, the competitors must form expectations of how the others react back on each of its own possible moves.

Equilibrium and Other Regular Orbits

If the system is defined as a recursive iterative map, i.e., sequence of moves, then the outcome, the orbit, can be a fixed point or equilibrium, as identified by Cournot in his original work already; a periodic oscillation, as discovered by Palander in 1939; a quasiperiodic orbit, i.e. one that almost repeats itself but never exactly; or chaotic, where the orbit is void of any identifiable pattern at all. No other possibilities exist for any dynamical system.

Obviously, the regular orbits, fixed points and periodic cycles (of not too long period), can be identified and learned by the competitors. In the case of a fixed point the situation is unproblematic. The competitors act assuming that equilibrium is maintained, and the outcome is exactly that. Nobody gets disappointed, so the whole idea of Cournot equilibrium is selfconsistent.

Throughout we will keep to the concept of Cournot equilibrium, though some authors switched to calling it Nash equilibrium, a much later concept, slightly more general, as it also applies to probabilistic settings, but in this work we skip such issues. Game theory, which has a long history from Benoulli back in the eighteenth century on, and which was launched as a tool for economic analysis by von Neumann and Morgenstern, may seem to be an adequate setting for oligopoly theory, but it keeps the theory in an equilibrium format, whereas we are concentrating on the dynamics of evolution.

A periodic orbit too, at least of sufficiently low period, can be learned by the competitors, but, as we will see, there is snag: Learning a periodic orbit and adapting to it generally changes the periodicity that is the outcome. It seems that there is no periodicity, if learned and adapted to, which produces that very periodicity as an outcome. This blows up the favourite concept of "rational expectations" in current economic theory.

It goes without saying that a chaotic or a quasiperiodic orbit, whose period is infinite, cannot at all be learned by the market agents, so, virtually, only the fixed point remains as a candidate for a dynamical oligopoly system where expectations can come out true. In all other cases, the formation of expectations remains a problem, and the competitors are bound to remain constantly disappointed.

Multiplicity of Attractors

Fixed points or equilibria can be unique or multiple, coexistent with each other and/or with other types of attracting orbits, such as periodic. To be quite exact, any linear model has just one unique attractor, but nonlinearity opens up for coexistence. Depending on the initial position of a dynamical system it can go to one or another attractor, the attracting regions, the basins, are well defined for each system, but they, as well as their boundaries, can be simple or very complicated, even fractal in design.

Linearity has been a popular first simplifying assumption in most sciences, but economics has kept to this longer than for instance physics, chemistry, or biology.

However, nonlinearities were introduced quite early in economics in this very area of investigation, for monopoly by Joan Robinson in 1933, and for duopoly by Tord Palander in 1939. It is seldom realized that piecewise linear models are *non*linear, and that for instance even an assumed linear demand function becomes piecewise linear due to the simple fact that prices and quantities cannot become negative.

The idea, introduced by Robinson, also used by Palander, was more complex than that. Robinson argued that, assuming linear demand functions for each group of consumers representing, for instance, a certain income class, market demand would display as a train of straight line segments of increasing elasticity as price is lowered. With a high price a commodity is a luxury reserved for consumers with high incomes; when price is lowered new and more numerous groups of consumers could afford it. The marginal revenue curve corresponding to such a demand curve would be a train of disjoint line segments with jumps between them. Hence any marginal cost function might intersect this train in several points, thus producing multiple coexistent local profit maxima for the monopolist.

Robinson assumes the monopolist to be conservative, if one local optimum has been found then the monopolist will stick to it, without comparing with the other local optima to find the global one. Hence, Robinson's theory remains static.

Palander transfers it to a market with interacting sellers, an oligopoly, and assumes that each seller always selects the global profit maximum. However, it depends on how much of the market the competitor takes; the global maximum shifts from one branch to another depending on this, so the reaction functions, representing the best reply to each assumed action of the competitor, come in several disjoint pieces. When checking for Cournot equilibria, i.e., intersections of the reaction functions for the competitors, one may then find several intersections, or none at all.

This may produce multiple attractors. For a duopoly Palander describes a case with two coexistent Cournot equilibria, also coexisting with a periodic oscillation. Palander presented his full argument in 1939, unfortunately in Swedish, so it has literally been buried. He made a presentation at the Cowles Commission in 1936, from which there exists a short printed statement in English, but it is difficult to judge how much of the argument was presented there. Some of it was reinvented under claims of originality as much as 20 years later.

The fact that Palander presented a full characterisation of coexistent attractors also makes his contribution the first global mathematical model ever presented in economics, even though the model suggested by Robinson has all features of a global model already. As she considers a monopolist's optimum a static framework is a natural choice; even though search mechanisms for the global optimum could be devised the construct would inevitably remain a bit construed. For Palander, considering duopoly, the stepwise adjustment between the competitors is necessary and so results in a fully dynamic model whose resulting orbits can be interesting objects of study.

The Robinson–Palander global system remained a solitaire for quite a long period, as mathematical economics was so focused on equilibrium and local stability. One may find different reasons for that: One is ideological, an inclination to harmony, rest, and optimality. Another is the scarcity of useful explicit functions that convincingly hold not just in small neighbourhoods, but over substantial regions of phase spaces where evolving orbits could be located.

Before leaving the issue of attractor multiplicity we should also mention Abraham Wald and his oligopoly model from 1936. It has not only multiple, but an infinite number of Cournot equilibria. Wald assumes smooth demand functions where an arc of a circle or a hyperbola fuses at tangency with linear segments on either side that project to the axes. Under his other assumptions this results in reaction functions for the competitors that coincide along a finite segment on which each and every point is a Cournot equilibrium. It is curious that Wald's contribution concerning imperfect markets is never cited, even if it is part of the article where the existence problem for general equilibrium, first attacked by Walras and finally solved by Arrow and Debreu, was dealt with and in that capacity is constantly listed as an important stepping stone in developing the proof.

Strategy Types

Once the focus is shifted from equilibria, unique or multiple, to the dynamic format we have to consider how expectations are formed and which strategies are chosen by the competitors. In the Cournot setting, there is no question of which action is best, provided one knows what the competitor(s) will do next.

The only problem is that *nobody knows that*. As mentioned, the most natural assumption in order to formulate a recursive system is to assume that each competitor assumes the other(s) to retain their previous moves. As the Cournot process evolves under the hypothesis that the agents assume their competitors to retain their last decisions, everybody will be constantly disappointed, because the assumption shows up not to hold true. Such expectations are now generally called "naive" or "myopic". Models based on such expectations may still produce interesting results, but, undeniably, they involve an element of inconsistency in modelling.

This is generally recognised. However, as mentioned above, the situation is aggravated by the fact, which is not generally recognized, that even if orbits are as simple as strictly periodic, learning these by the actors and adapting to them, is impossible as this alters the periodicity. As other possible orbits, chaotic and quasiperiodic, are too complicated to learn, this leaves us with the fixed points as the only possible orbits where the competitors are not constantly deceived in their expectations. This undeniably is a problem. However, one could say that in an actual oligopoly market expectations are somehow formed and actions are taken. If modellers cannot formulate a consistent system where expectations come out true and nobody is disappointed, why should the actual competitors be better? After all the naive expectations are the simplest for both modeller and real agent.

Different amendments have been made, such as assuming that the agents apply some measure of conservatism; not jumping from their previous moves straight to the new calculated best replies, but only part of the way. Then adaptation weights, between zero and unity, are introduced for retaining the previous action and for moving to the calculated best reply. The results of this simple change often turn out as quite interesting because for a duopoly the system then really becomes two-dimensional. If the duopolists react only to the competitor's last move, then a composition of two steps results in a second order system for each competitor alone, and the events on the market just become a combination of such independent iterations. Like the naive adjustment, the adaptive one might be a reasonable modelling assumption. Even if profits do not always attain their possible maximum values, they need not fall so much short of these over an extended period that the competitors have reason to be really disappointed. Economics is usually concerned with the exact location of optima, but very seldom with the sharpness of the optimal goal realisation over alternatives.

We should add here that the classical writers, for instance Palander, made it an important distinction whether the duopolists react simultaneously or in sequence. Now one knows that such a distinction is immaterial for the outcome of a system. Simultaneous adjustment can produce sequential orbits. The only difference is that only in the latter case are the points of the orbit located on the reaction functions. Depending on initial conditions a given system can go to such a sequential orbit or not.

Stackelberg Leadership

Above the possibility of learning actually outcoming orbits was discussed, but there are other things that the competitors might learn in order to make a better performance. A remarkable idea was suggested by Heinrich von

Stackelberg in 1934: each competitor, for simplicity duopolist, might learn the reaction function of the other. In this way, all reactions of the competitors can be taken in account. The competitor called market leader can calculate its best possible solution, and choose Stackelberg leadership. As Stackelberg realized, this works, if and only if the other competitor indeed adheres to its reaction function according to Cournot. If not then the solution does not work.

Whereas Cournot's theory came both as a definition of the Cournot equilibrium state, and as an iterative system, Stackelberg's only came in the equilibrium format. The leader calculates the maximum profit state which is just a point, and the follower who acts according to the Cournot reaction function then also picks just one point on this function. Obviously, either firm could become a leader. If the other adheres to the proper Cournot reaction function, then the Stackelberg equilibrium is a feasible solution. There are thus two such Stackelberg equilibria. If both try to become leaders at the same time, both are deceived, and the solution is infeasible. If both act as followers, then we are back at the Cournot equilibrium, which like the two Stackelberg leader/follower equilibria, is a feasible solution. As Cournot equilibrium remains, it may thus seem that Stackelberg generalized Cournot's theory, but this is only partially true, more precisely as long as equilibria are concerned.

In view of dynamics there is no clue in Stackelberg's framework. As it departs from Cournot's theory, an attempt to bring Stackelberg's idea into a dynamic setting would most naturally try the Cournot dynamic framework. One may assume that during a Cournot process, preferably depending on some endogenous clause, any competitor might occasionally try to become a Stackelberg leader.

If one does, one inevitably runs into a paradox: Suppose one competitor at some time tries Stackelberg leadership action. Further, suppose the competitor indeed reacts according to the proper Cournot reaction function, quite as assumed as prerequisite for calculating Stackelberg leadership. What is then the best next reply for the firm that tried leadership; keep to the leadership or revert to Cournot action?

Paradoxically, it is always better to revert to Cournot action, just because the reaction functions are defined the way they are, as the best actions under any move by the competitor.

It is proven that Stackelberg *equilibrium* profit is always higher than Cournot *equilibrium* profit. But it is *not* true that choosing Stackelberg action during an evolving dynamic process yields higher profit than Cournot action, not even when the system actually is in a feasible Stackelberg equilibrium point.

Of course, reverting to Cournot action reintroduces the Cournot dynamics leading to Cournot equilibrium, and then in a longer perspective it would have been better for the leader to insist on leadership action. But the paradox remains; in the process, for temporary profit, it is always better for a leader to revert to Cournot action in the next move.

This leads to a more general question concerning Stackelberg's modelling: If a would be leader believes the competitor to keep to follower action, why not try to launch a series of moves changing over time, dragging the follower along, and hoping to get an even higher average profit over time than that which results from a Stackelberg equilibrium, constant over time. Stackelberg never thought along these lines, as he, like Robinson, kept to the equilibrium format.

Stability of Cournot Equilibrium

An equilibrium, unique, or one among other attractors, when placed in a dynamic setting, is not interesting if it is not stable. Using linear demand functions and constant marginal costs, one inevitably gets reaction functions with a constant slope of minus one half. This is so because when price decreases linearly with market supply, revenue comes to contain a squared term of the firm's own supply. The other supplies enter as products with this own supply. Differentiation, to get marginal revenue, then results in an expression where the term containing the firm's own supply is doubled, whereas those of the other firms are not.

Solving for the firm's own supply in terms of those of the competitors one divides by two, which results in a constant slope of minus one half. This also holds for the Robinson–Palander case. In a duopoly, with two reaction curves, their intersection, i.e., the Cournot equilibrium point, is always stable. Note that as long as marginal costs are constant the cost side contributes nothing to complicate the issue. Differentiation of the cost terms then just results in constants. It is, by the way, surprising how frequent the use of constant, or even zero marginal cost was. Palander used it in his first example, and so did Wald, Hotelling, and many others.

Hence not only a duopoly with linear demand and constant marginal cost with one unique Cournot equilibrium, but also Palander's two concurrent Cournot equilibria are stable with each its proper basin of attraction.

The question then arises what happens if the number of competitors grows. Theocharis in a short article 1959 is assumed to have originated the idea that if the number of competitors grows beyond three, then the Cournot equilibrium is destabilised. However, this was pointed out by Palander already 20 years earlier, in the very paper already discussed.

It is easy to see, as the Jacobian matrix is composed by off-diagonal constants equal to one half, that there is one eigenvalue that exceeds the unit stability interval if there are more than three competitors.

This discovery has a certain relevance for the relation between oligopoly and perfect competition. Adding new competitors lowers the market price towards marginal cost, but if at the same time equilibrium is destabilised due to this increasing competition, how likely is it then that an oligopoly transforms into a perfect competition equilibrium?

As it turns out that the same destabilisation occurs also with other reasonable demand functions when the competitors produce with constant marginal cost, the natural question arises whether this latter assumption is a reasonable one. Constant marginal cost means production under constant returns to scale, but a firm producing under constant returns is in principle infinitely large as it under any differential between perfect market price and marginal cost can enlarge its profits to infinity just through increasing the scale of operation.

That adding such *infinite sized* firms destabilises equilibrium should neither be surprising nor relevant to the issue. Economists discussing this issue implicitly take that the comparison is between cases of *few large* oligopolists versus *numerous small* perfect competitors. This cannot be formulated without modelling large and small, which seems impossible without introducing decreasing returns and capacity limits. In this way the relevance of the problem attributed to Theocharis and stated by Palander much earlier, seems to fade away. As we will see in the sequel such capacity limits can be introduced in several ways. An issue related to studying competition among few large and many small firms is studying how competition may evolve *endogenously* in a branch over time when new firms enter and old ones exit or are reconstructed through renewal of worn out capital.

Exotic Dynamics

In a remarkable article in 1978 David Rand suggested oligopoly theory as an obvious candidate for producing complex dynamics, not only low period orbits such as studied by, for instance Palander, but quasiperiodic and even chaotic ones. Given instability of Cournot equilibrium is focused, this is a natural next step.

Rand did not suggest any piecewise linear model, but one where the reaction functions of duopolists were smooth unimodal curves with maxima, having four different intersections, i.e., possible coexistent Cournot equilibria, and then he studied parameter dependent routes to chaos along the lines used in current chaos studies, such as the discrete logistic iteration.

The problem was that Rand could not point out where such unimodal reaction functions came from. Until this day nobody has been able to formulate substantial economic assumptions which lead to such reaction functions as his.

A general comment on the difficulty to find acceptable assumptions such that lead to consistent and tractable oligopoly models is in place here. Economic modellers would always like to find consistent combinations of assumptions based on accepted economic principles. So, in order to derive the reaction functions, one would like to know which types of demand and cost functions might lead to reaction functions with desired properties. It goes without saying that such demand and cost functions must make sense in terms of economics intuition. Seen from the other end, starting from acceptable demand and cost functions, it would be desirable to derive nice closed form reaction functions.

One would further want to be able to calculate the coordinates of the Cournot equilibria. This is a formidable programme, and it must be admitted that it is very difficult to find demand and cost functions which allow for this analytic programme. We would like to derive these building blocks, demand and cost functions, from the microeconomics of consumers maximizing utility functions and firms minimizing production costs using reasonable production functions. This does not make the programme easier.

A related general problem is the need for dynamic analysis to have *global* functions to work with which hold over substantial regions of the phase spaces. Economics abounds of representations of the typical facts in terms of graphs and tables, but, as the focus for so long has been on equilibria and their local stability, few useful global functions appear in the literature.

The exception seems to be the functions introduced in empirical production studies, where production functions, such as the Cobb–Douglas, the CES, and the trans-log, were fitted to time series or panel data, and which hence needed to hold over substantial periods or over wide branch structures. The fit of the estimates, naturally, was never perfect, even though as many different parameters as could be accommodated were included, but it was good enough for these functions to become integrated also in theoretical production studies. From there, they were very soon transferred also to the theory of the consumers.

Though crude, these functions are useful for oligopoly theory. A Cobb–Douglas utility function allows for deriving iso-elastic demand functions, easily aggregated over individuals. Such demand functions further allow for deriving simple reaction functions, which come as close to the Rand ideas as possible, and display at least some of the phenomena he wanted to derive. Likewise, one can use the CES functions with fixed capital to arrive at cost functions with built in capacity limits such as wanted above.

Heterogeneous Commodities

It is now time to return to Bertrand's criticism of Cournot. It mainly concerns Cournot's choice to assume that the competitors compete through fixing their quantities of supply in the market, whereas the resulting price would then result through some unspecified market clearing mechanism. In the case of a homogenous good in a market with few competitors, it might actually seem to be more natural to compete through choosing supply prices. A problem arises in the case of homogenous commodities: Through undercutting the competitors' prices even infinitesimally would cut out the competitors, and the firm doing this could take the whole market.

As cited from Schumpeter above, Bertrand's argument was very confused and very aggressive, and was articulated by Edgeworth a few years later, in 1897.

The solution is to assume that the commodity supplied by different suppliers is conceived as slightly different by the consumers, either because of actual physical differences, or because the consumers just have different preferences for particular suppliers. To this field belong the studies of price discrimination, through identifying different consumer groups with different demand elasticities and making them have preference for one particular product, among essentially identical types, even if only wrapped up in some special way. This field of study culminated in Chamberlin's theory of monopolistic competition dating from 1933.

The problem with such modelling is that the idea of close substitutes was never formalized in a way to produce any convincing demand functions. Sir John Hicks (1946), for instance, made a lot of advance in classifying substitutes and complements in gross sense, including income effects, or net sense, based on substitution effects alone, but all these classifications were based on *local* characteristics in some given market situation, and nobody has as yet proposed a convincing demand theory for substitutes and complements useful for modelling a global dynamic process.

A utility function of Cobb–Douglas type, for instance, results in demand functions where demand for the commodities is independent in terms of gross substitutivity, both for the individual consumer and for the entire market. Maybe, it is for this reason that Cournot's original model, despite its shortcoming, kept its position as the main frame for dynamic oligopoly over a period that soon stretches over two centuries.

The Hotelling Paradox

More suitable for modelling than some general difference in the perception of similar commodities is the case where one takes a physically homogenous commodity, perceived as such by the consumers, but lets it be supplied in different locations. Consumers have the choice of one supplier or another, but if they are located at appreciable distance in space, then the suppliers become local monopolists, surrounded by proper market areas separated by boundaries where the supply prices, so called mill prices, accrued with transportation costs break even.

Then price differentials between different supplies of the same commodity can be set up and maintained, though there is competition at the boundaries. When one firm raises its local monopoly price, its market area inevitably decreases; it can charge a higher price but sell less and has to struck some optimal balance between these two effects.

Hotelling's famous paradox of 1929 arises from the fact that he also took the choice of location, not as fixed, but optional for the competitors. Hotelling's case is chosen as the one-dimensional line.

The reader, by the way, may be surprised how many attempts by economists to treat phenomena in the geographical two dimensional plane, just stop after discussing one dimension, explained to be a simplifying introductory step. In one dimension the market areas are just intervals and the boundary consists of a point. The case of phenomena with circular symmetry is the same, seemingly two-dimensional but substantially just one-dimensional.

When taking the entire region as a *fixed interval*, then paradoxically the firms cluster in the midpoint of this interval, in stead of locating in the centres of each one half-interval, which would be socially optimal in terms of minimizing the total need for transportation. It should be noted that the paradox does not arise if the interval is not bounded, but, for instance the perimeter of a circle. The reason for clustering being to obtain as large a market area as possible in order to maximize total market sales, this does not happen on a closed circle.

The problem is that there is not only the question of trying to cut out the competitor, as in the Cournot case when price is used for competition, but that the local optima in separate intervals when both mill price and location choice are optimised become indeterminate.

This naturally was considered so much a problem that different remedies were sought. One of the most construed was the proposition by d'Aspremont et al. (1979) to make transportation cost dependent on distance, not linearly, but quadratically. Even though it only applies with realism to something like a fast decaying radioactive substance, there have been a sizeable number of authors that used this assumption.

The needed higher order for the optimization to obtain determinate optima can, however, be easily obtained through the more reasonable assumption of elastic demand. Hotelling himself was conscious of an obvious inconsistency in his model: Whereas price including transportation is relevant for the consumers for establishing market boundaries, as they would not buy from suppliers to whom they have to pay more, the quantity of demand is assumed to be independent of price, as the consumers located on one distance unit of the interval always buy one unit of the commodity, no more and no less, independent of its price. Only in this way does the total sales become proportional to the extent of the market area. In this way calculations are simplified, but at the same time no local optima for plant locations and mill prices can be found.

Hotelling himself in a verbal comment conjectured that the indeterminacy will be removed if elastic demand is introduced. A first attempt at doing this was made in an ingenious graphic analysis by Lerner and Singer (1937), where they introduced a reservation price; up to this consumers would buy one unit of the commodity as assumed by Hotelling, but if the actual cost exceeded the reservation price, they would buy nothing. It was seen that this removed the paradox. A few years later, Smithies (1941a, b) advocated the use of a continuously decreasing demand function, such as the linear, but curiously enough claimed the analysis to be too complicated mathematically to be carried out. In fact the integrations needed are quite straightforward in contrast to Smithies's claims.

Agenda

This attempt at describing the scenery in dynamic oligopoly theory as it emerged in the approximately 100 years 1838–1941, indicates quite a number of loose ends which will be taken up in the sequel. After that period not much happened for quite a long period. There seems to be a twin reason for this.

One reason is the dominance of general equilibrium theory from about 1950, with its focus on existence of equilibria, convex structures and uniqueness, where even dynamic orbits were hidden in the index numbers of intertemporal equilibrium points, with period dating intemingled with the detailed physical characteristics of commodities. Main focus further was on the system of competitive markets, where market imperfections have little room.

Another reason is that oligopoly, ever since the publication of von John Neumann's and Oscar Morgenstern's great monograph in 1943 was regarded as a proper field for applications of the theory for games of hazard, such as betting or playing poker. In the search for probabilistic strategies to play one is, however, always looking for sustainable equilibrium constellations. There undeniably exists a sizeable and interesting literature about such game theoretic oligopoly models, but, within this frame, it was not suitable to continue developing the classical theories of oligopoly and imperfect markets as described above.

This leaves a lot of its loose ends unexplored, to the benefit of present day modellers who set out to explore more exotic phenomena of oligopoly dynamics.

The sequel is kept to the strictly dynamic outlook, avoiding all reference to game theory and its concepts. In recent literature on dynamic oligopoly there is frequent use of the concept of oligopoly *games*. Simple old Cournot equilibria are renamed *Nash equilibria*, though nothing probabilistic is involved. One speaks of *Markov perfect equilibria* in connection with periodic orbits with no touch of equilibria to them, and the like. Perhaps the reason is to ripe some prestige from current game theory, but in the opinion of the present author this may be confusing, and will hence be avoided in the sequel, which sticks to the classical theory as developed around 1940, though using the means of analysis that modern dynamics offers.

Even if most of the queries and loose ends listed in this chapter are taken up for further discussion, it must be stressed that the present work is not a systematic and comprehensive study of market imperfections. Studies aiming at this, as for instance Stackelberg's, normally treat all relevant issues, such as monopsony and oligopsony, bilateral monopoly, and price discrimination policies. In the sequel focus will be kept on the case of few sellers competing in a market with atomistic consumers. Oligopsony is an interesting phenomenon in the raw materials markets, but the analysis is just the mirror image of the analysis of oligopoly, which one meets everywhere in a modern economy.

Assumptions

Demand

Above, there was mention of the scarcity of closed form global functions in economics available for use in dynamic systems where orbits can take routes through substantial regions of the phase space. However, we constantly mentioned demand and cost functions, and there is not so much more we need.

As for demand functions, linear ones have appeared frequently in previous discussions, and they will also be used to discuss some of the topics dealt with, such as the Palander–Theocharis stability problem, or the Hotelling model with elastic demand as proposed by Smithies. Also the Robinson–Palander idea of kinked demand functions will be used as it gives such interesting focus to coexistence of attractors. After all, the kinked train of line segments is a logical consequence of aggregation of individual demand functions for an entire market.

In addition the isoelastic demand function, a hyperbola where price and market demand are reciprocal, will be used. It has certain advantages. Unlike the linear example it results from an acceptable utility function; the Cobb–Douglas which always makes income be split in fixed income shares for the different commodities. Further, also unlike the linear case, the individual functions aggregate to a market function of the same form as the individual ones. Despite its nonlinear nature this smooth function behaves well as one can both solve for the closed form reaction functions and for the coordinates of the Cournot equilibrium point. The model comes as close as possible to Rand's conjecture and lets one explore some exotic dynamics

Finally, in dealing with Stackelberg, a demand function where price is a an exponential of the negative of the total quantity supplied on the market is used. It is somewhat ad hoc, quite as the linear, but it has the historical advantage of having been used by Stackelberg in his original contribution.

Cost

As for cost, the assumption of constant marginal cost will be used quite as in previous literature, though the zero cost which was quite popular in early writings will be avoided.

In the context of modelling small and large firms, a version of the CES function in stead will be employed, where capital is fixed during investment periods, and puts a capacity limit for production, even though the remaining inputs are assumed to have full variability. More than so will not be needed.

References

d'Aspremont, C., Gabszewitsc, J. J., & Thisse, J. F. (1979). On Hotelling's "stability in competition". *Econometrica, 47*, 1145–1150.

Bertrand, J. (1883). Théorie mathématique de la richesse sociale. *Journal des Savants, 48*, 499–508.

Chamberlin, E. H. (1933). *The economics of monopolistic competition* Cambridge. MA Harvard University Press:
Cournot, A. (1838). *Récherces sur les principes mathématiques de la théorie des richesses.* Paris: Donod.
Edgeworth, F. Y. (1897). La teoria pura del monopolio. *Giornale degli Economisti, 15*, 13–31.
Fellner, W. (1949). *Competition Among the Few.* (Alfred A. Knopf, New York)
Hicks, J. R. (1946). *Value and Capital.* Oxford: Oxford University Press.
Hotelling, H. (1929). Stability in competition. *Economic Journal, 39*, 41–57
Lerner, A. P., & Singer, H. W. (1937). Some notes on duopoly and spatial competition. *Journal of Political Economy, 45*, 145–186.
von Neumann, J., & Morgenstern, O. (1943). *Theory of Games and Economic Behavior.* Princeton: Princeton University Press.
Palander, T. F. (1936). "Instability in competition between two sellers". Abstracts of papers presented at the research conference on economics and statistics held by the Cowles Commission at Colorado College, *Colorado College Publications*, General Series No. 208, Studies Series No. 21.
Palander, T. F. (1939). Konkurrens och marknadsjämvikt vid duopol och oligopol. *Ekonomisk Tidskrift, 41*, 124–145, 222–250.
Rand, D. (1978). Exotic phenomena in games and duopoly models. *Journal of Mathematical Economics, 5*, 173–184.
Robinson, J. (1933). *The Economics of Imperfect Competition.* London: Macmillan.
Schumpeter, J. A. (1954). *History of Economic Analysis.* London: George Allen & Unwin Ltd.,
Smithies, A. (1941). Optimum location in spatial competition. *Journal of Political Economy, 49*, 423–439.
Smithies, A. (1941). Monopolistic price policy in a spatial market. *Econometrica, 9*, 63–73.
von Stackelberg, H. (1934). *Marktform und Gleichgewicht.* Berlin: Julius Springer.
Theocharis, R. D. (1959). On the stability of the Cournot solution on the oligopoly problem. *Review of Economic Studies, 27*, 133–134.
Wald, A. (1936). Über einige Gleichungssysteme der mathematishen Ökonomie. *Zeitschrift für Nationaökonomie, 7*, 637–670.

2 Cournot Oligopoly

The Cournot Model

As mentioned in Chap. 1, Cournot's oligopoly model was one of the first mathematical models proposed in the field of economics See Cournot (1838). It addresses the functioning of a market with numerous atomistic demanders versus few relatively large suppliers. This implies that all the suppliers influence market price appreciably, and hence, like monopolists, take account of the demand function of the consumers on the market in order to calculate their best moves. As a rule, demand is a decreasing function of price. In equilibrium demand equals supply, and one can also speak of the inverse demand function which states how market price depends on supply. In the case of Cournot, it is most convenient to speak of this inverse demand function,

$$p = f(Q), \tag{2.1}$$

where p denotes price, Q denotes market supply, and $f'(Q) < 0$.

Cournot takes the quantity of supply for each oligopolist as the proper decision variable, so the price somehow results through a not specified market clearing mechanism. This was later the goal for criticism, as a few big firms would rather set supply prices than quantities supplied; and this would open up for the possibility of cutting out competitors in case the commodity marketed is conceived as homogenous by the consumers.

Supposing there are n suppliers whose individual supplies are denoted q_i, market supply becomes,

$$Q = \sum_{i=1}^{i=n} q_i \ . \tag{2.2}$$

T. Puu, *Oligopoly: Old Ends – New Means*, DOI 10.1007/978-3-642-15964-0_2,

There is a point in defining residual supply,

$$Q_i = Q - q_i, \tag{2.3}$$

the supply of all the other firms, which is not under the control of the ith firm.
The profits, revenue minus cost for the ith firm now become

$$\Pi_i = f(q_i + Q_i)q_i - C_i(q_i), \tag{2.4}$$

where $C_i(q_i)$ is the cost function. Differentiating partially with respect to q_i, results in the first order condition,

$$f(q_i + Q_i) + f'(q_i + Q_i)q_i = C_i'(q_i). \tag{2.5}$$

Unlike the case of monopoly, the optimum condition does not determine the value of q_i; the solution depends on Q_i, the residual supply by the competitors about which the ith firm has no certain knowledge. It can only have more or less sophisticated expectations, based on past experience, and calculate the best decisions under each such expected Q_i.

The outcome is the reaction function,

$$q_i = \phi_i(Q_i), \tag{2.6}$$

which reduces the optimality condition to an identity, i.e.,

$$f(\phi_i(Q_i) + Q_i) + f'(\phi_i(Q_i) + Q_i)\phi_i(Q_i) \equiv C_i'(\phi_i(Q_i)). \tag{2.7}$$

To find such reaction functions $\phi_i(Q_i)$ in nice closed form would be desirable for any modeller. Unfortunately, there are very few demand and cost functions that allow one to do so. Next follow a few examples where this programme can be accomplished, i.e., the traditional case of a linear demand function, and the case of an iso-elastic, or hyperbola shaped demand function.

Example 1: Linear Demand

Suppose we have the inverse demand function

$$p = a - bQ, \tag{2.8}$$

where a, b are two positive constants. Further, assume the cost functions are

$$C_i = c_i q_i,$$

where c_i are constant marginal, equal to average variable costs.

Then the profit of the ith firm becomes,

$$\Pi_i = (a - b(q_i + Q_i))q_i - c_i q_i = (a - c_i - bQ_i)q_i - bq_i^2, \tag{2.9}$$

and the maximum condition,

$$a - c_i - bQ_i = 2bq_i. \tag{2.10}$$

Solving, the reaction function is readily obtained,

$$q_i = \frac{a - c_i}{2b} - \frac{1}{2} Q_i. \tag{2.11}$$

These reaction functions are straight lines with the constant slope $-\frac{1}{2}$. Obviously we must have $a > c_i$, i.e., the maximum price obtainable must exceed the unit cost, otherwise the firm could not obtain any profit. Further, the reaction function would return a negative output q_i, unless

$$Q_i < \frac{a - c_i}{b} \tag{2.12}$$

holds. As this is meaningless, any negative q_i, would be replaced by zero, which means that if the competitors supply too much, then the firm drops out.

To get a complete picture it is better to explicitly check for nonnegativity of profits.[1] Substituting the reaction function in the profit expression one obtains

$$\Pi_i = \left(a - c_i - bQ_i\right)\left(\frac{a-c_i}{2b} - \frac{1}{2}Q_i\right) - b\left(\frac{a-c_i}{2b} - \frac{1}{2}Q_i\right)^2, \tag{2.13}$$

which equals

$$\Pi_i = \frac{b}{4}\left(\frac{a-c_i}{b} - Q_i\right)^2. \tag{2.14}$$

It may seem that this expression is always nonnegative, but when the reaction q_i is negative, this is due to the fact that negative costs outweigh negative revenues, which has no factual meaning. Accordingly, considerations of profits add nothing new to the constraint for a positive reaction already stated, so the complete reaction function reads

$$q_i = \begin{cases} \dfrac{a-c_i}{2b} - \dfrac{1}{2}Q_i, & Q_i < \dfrac{a-c_i}{b} \\ 0, & Q_i \geq \dfrac{a-c_i}{b} \end{cases}. \tag{2.15}$$

Note that the resulting reaction function is piecewise linear, i.e., nonlinear, even in this simplest case with a linear demand function and constant unit costs. In setting up these reaction functions, all the steps needed to be considered at setting up reaction functions in further examples were already encountered.

---

[1] Observe that the non-negativity constraint refers to calculated profits based on the competitors' expected moves. If they turn out wrong, then in a dynamic process the firms may still experience actual negative profits.

Example 2: Isoelastic Demand

Linear demand functions are just simple first approximations. If one assumes individual consumers or groups of consumers to have different linear demand functions, then market demand inevitably results as a broken train of line segments, as proposed by Robinson (1933). Used in oligopoly models it results in the interesting phenomena described by Palander (1936, 1939), but the analysis becomes more messy than is suitable for this simple exemplification, so we defer the analysis of this case to a later chapter.

In stead we now assume that the consumers maximize utility functions of a Cobb–Douglas type, a product of fractional powers of the quantities of different commodities consumed.[2] Without loss of generality such utility functions can be rescaled so that the sum of these power fractions equals unity. Maximizing such utility functions under given budget constraints, then result in income being split in budget shares where the power fractions from the utility function provide the weights. Fixing budget shares of a given income means that quantity demanded and price are reciprocal. See Puu (1991, 2004).

Passing from the individual demand functions to the market, as all individual functions are of the same form results in market demand as well being reciprocal to price, the numerator just being the sum of all the individual budget shares.

As the measurement unit for the commodity is optional, one can choose it in such a way as to normalize the numerator to unity, and thus obtain the inverse demand function

$$p = \frac{1}{Q} \tag{2.16}$$

---

[2] Assume a utility function $U = q_1^{\alpha_1} \cdot q_2^{\alpha_2} \cdot \ldots q_n^{\alpha_n}$ for a representative consumer, where the q_i for the moment represent quantities of *different* commodities consumed, n is the number of commodities, and α_i are some constants, such that $\alpha_1 + \alpha_2 + \ldots \alpha_n = 1$ holds. The consumer, maximizing utility under the budget constraint $y = p_1 q_1 + p_2 q_2 + \ldots p_n q_n$, then chooses $p_i q_i = \alpha_i y$, which is the same as $q = \alpha y / p$, where we drop the indices as only one commodity is of interest. Summing over all consumers $Q = \sum \alpha y / p$. The consumer identification indices for the α :s and y:s have been suppressed in order not to overload notation, but they are implicit, and can be different for all the consumers. Market price p, on the contrary, is the same for all consumers.

The profits of a representative firm now become

$$\Pi_i = \frac{q_i}{q_i + Q_i} - c_i q_i, \tag{2.17}$$

again assuming constant unit costs, and the profit maximum condition reads,

$$\frac{Q_i}{\left(q_i + Q_i\right)^2} = c_i . \tag{2.18}$$

As the numerator, residual demand, as well the right hand side, unit cost, are positive, one can take roots, and solve for the simple reaction function,

$$q_i = \sqrt{\frac{Q_i}{c_i}} - Q_i \tag{2.19}$$

This function has a leading square root term of Q_i, and an eventually dominant negative linear term, so its shape is unimodal, starting in the origin, increasing to a maximum, and then decreasing to zero.

The reaction q_i remains positive as long as

$$Q_i < \frac{1}{c_i} \tag{2.20}$$

holds. If not, then again the negative outcome will have to be replaced by zero.

Accordingly, the complete reaction function reads

$$q_i = \begin{cases} \sqrt{\dfrac{Q_i}{c_i}} - Q_i, & Q_i < \dfrac{1}{c_i} \\ 0, & Q_i \geq \dfrac{1}{c_i} \end{cases} . \tag{2.21}$$

One should check profits for nonnegativity. Substituting the nonzero branch of the reaction function in the profit expression yields

$$\Pi_i = \frac{\sqrt{\frac{Q_i}{c_i}} - Q_i}{\sqrt{\frac{Q_i}{c_i}}} - c_i\left(\sqrt{\frac{Q_i}{c_i}} - Q_i\right) = \left(1 - \sqrt{c_i Q_i}\right)^2 . \tag{2.22}$$

Again profit is nonnegative, but again, when the reaction is negative, it is due to negative costs outweighing negative revenues, which is nonsense. Even this time we, however, do not need to introduce any more constraints than already represented by the zero branch condition.

This case with isoelastic demand is quite useful for displaying some of the interesting dynamics possibilities of oligopoly theory, but it has its limitations: In particular, it is not suitable for the discussion of monopoly or collusion. This is because if there is just one firm, the monopolist, then it receives a constant (unit) revenue, $pQ = 1$. On the other hand, production costs increase with output, and all variable costs may be avoided if the monopolist produces nothing. It then sells this nothing at an infinite price, as may be seen from the demand function. In terms of substance this is nonsense, so one just cannot use the model for the monopoly case.

In dynamic oligopoly, the presence of other competitors as a rule keeps the system from landing in zero output for all firms, a case of implicit collusion, which has the same characteristics as monopoly. Further the derived reaction functions all have infinite slope at the origin, so under any perturbation, the system will never stick there. But, in a complete characterization of possible orbits, there remain degenerate orbits which even have their basins of attraction and are stable in a weak Milnor sense. See Tramontana et al. (2010).

Cournot Equilibria

Next consider the equilibria of the general Cournot model. The optimum conditions

$$f(Q) - f'(Q)q_i = C'(q_i) \tag{2.23}$$

with the definition $Q = \sum_{i=1}^{i=n} q_i$ added provide a set of $(n + 1)$ equations in the same number of unknowns, q_i and Q. The solution(s) provide the coordinates of the Cournot equilibria. This also means that all the reaction functions $q_i = \phi_i(Q_i)$ with definition $Q_i = \sum_{j \neq i} q_j$ are satisfied. When the competitors are so few that one can display the phase space graphically, for instance two, the duopoly, then the Cournot equilibria are intersections of the reaction curves. With three competitors, already difficult to visualize, they are intersections in three-space of three surfaces.

Example 1: Linear Demand

In the two exemplifying cases it is possible to calculate the coordinates for the Cournot point. Starting with the linear, the optimality conditions read

$$q_i = \frac{a - c_i}{2b} - \frac{1}{2} Q_i . \tag{2.24}$$

Subtracting $\frac{1}{2} q_i$ from both sides, and multiplying with 2,

$$q_i = \frac{a - c_i}{b} - Q , \tag{2.25}$$

Then, taking the sum over index *i*,

$$Q = \frac{na - \sum_{i=1}^{i=n} c_i}{b} - nQ \tag{2.26}$$

is obtained, which solves for

$$Q = \frac{n}{n+1} \frac{a - c}{b} , \tag{2.27}$$

where

$$c = \frac{1}{n}\sum_{i=1}^{i=n} c_i \tag{2.28}$$

denotes average unit cost for the competitors. Substituting for total supply in the Cournot equilibrium point, one gets

$$q_i = \frac{a - c_i}{b} - \frac{n}{n+1}\frac{a-c}{b} \tag{2.29}$$

for the supplies of the individual firms.

It is also easy to calculate the equilibrium market price, through substituting for market supply Q in the inverse demand function. Hence

$$p = a - bQ = \frac{1}{n+1}a + \frac{n}{n+1}c\,. \tag{2.30}$$

This is a well known result from work with linear demand functions: Equilibrium price equals a weighted average of maximum price a, and unit production cost c. The more firms that stay active in the oligopoly, the more weight has the second term and the less has the first. In monopoly, price lands halfway between maximum price and unit cost, in duopoly the weight for the cost term is two thirds and one third for maximum price. With an increasing number of competitors the maximum price term tends to lose all importance, and one approaches marginal cost pricing, as prescribed in perfect competition.

To consider which firms will stay in the oligopoly market, one could either check the Cournot equilibrium supply coordinate for positivity, or check that unit cost be lower than market price, i.e.,

$$c_i < \frac{1}{n+1}a + \frac{n}{n+1}c \tag{2.31}$$

It must be kept in mind that the definition of c contains also the value of c_i. See Canovas et al. (2009). To squeeze out some more information, suppose the firms are numbered after increasing unit costs, i.e.,

$$c_1 < c_2 < \ldots c_n, \tag{2.32}$$

which can be done without loss of generality. Then the conditions

$$c_1 < a, \quad c_2 < \frac{a + c_1}{2}, \quad c_3 < \frac{a + c_1 + c_2}{3}, \quad \& \tag{2.33}$$

or, in general,

$$c_i < \frac{1}{i}\left(a + \sum_{j=1}^{j=i-1} c_j\right), \tag{2.34}$$

are obtained from (2.31).

Hence, how many competitors that may be accommodated in a Cournot equilibrium depends on the precise cost structure in that oligopoly branch. In the special case where all the firms are identical, i.e., $c_i = c$, it is sufficient that $c_i = c < a$. If, on the other hand, the cost situation is very unequal, for instance, $c_2 > (a + c_1)/2$, then only a monopoly of the first firm can make positive profits, if $c_3 > (a + c_1 + c_2)/3$, then a duopoly can persist, etc.

Example 2: Isoelastic Demand

Also for the second example it is possible to obtain the closed form expressions for the Cournot point coordinates and the rest of the results just discussed. Recall the main branches (2.19) of the reaction functions,

$$q_i = \sqrt{\frac{Q_i}{c_i}} - Q_i. \tag{2.35}$$

Adding Q_i to both sides,

$$Q = \sqrt{\frac{Q_i}{c_i}}, \tag{2.36}$$

or, taking squares,

$$Q^2 = \frac{Q_i}{c_i}, \tag{2.37}$$

is obtained.

Multiplying by c_i, recalling that $Q_i = Q - q_i$, and reorganizing slightly,

$$q_i = Q - c_i Q^2, \tag{2.38}$$

Taking the sum over *i*, one gets

$$Q = nQ - ncQ^2, \tag{2.39}$$

where, like in the previous case,

$$c = \frac{1}{n} \sum_{i=1}^{i=n} c_i \tag{2.40}$$

denotes average unit cost.

As $Q > 0$, it is permitted to cancel one power in equation (2.39), so

$$Q = \frac{n-1}{n} \frac{1}{c}, \tag{2.41}$$

and, consequently,

$$q_i = \frac{n-1}{n} \frac{1}{c} \left(1 - \frac{n-1}{n} \frac{c_i}{c} \right) = \frac{n(n-1)c - (n-1)^2 c_i}{(nc)^2}. \tag{2.42}$$

Several facts are worth being noted. As indicated in the introduction, the model is not suitable for discussing monopoly. The conditions (2.41) and (2.42) result in zero output when $n = 1$. One can also easily calculate oligopoly price in the Cournot equilibrium:

$$p = \frac{n}{n-1} c . \tag{2.43}$$

Disregarding again the degenerate monopoly case where price is infinite, equilibrium price is twice the unit cost for a duopoly, one and a half times cost in a triopoly, and so forth. Again, as the number of competitors increases, one approaches marginal cost pricing and the case of perfect competition.

As for the question which firms may obtain positive profits in the Cournot equilibrium, one can again either consider positivity of the reaction, or that unit cost be lower than market price, i.e.,

$$c_i < \frac{n}{n-1} c . \tag{2.44}$$

Again, it should be noted that c is the average of all the unit costs, including that on the left, and again, one can assume the firms to be numbered in order of increasing unit costs. In this way, ignoring the degenerate case of monopoly, one has $c_2 < c_1 + c_2$ for duopoly, which is always fulfilled with positive unit costs, $2c_3 < c_1 + c_2 + c_3$ for triopoly, which is the same as $c_3 < c_1 + c_2$, or, in general.

$$c_i < \frac{1}{i-2} \sum_{j=1}^{j=i-1} c_j \tag{2.45}$$

Hence, for two competitors in a Cournot oligopoly, there are no constraints for the costs; for three, the unit cost of the third must, however, not exceed the sum of the costs of the two firms with the lowest costs. A fourth firm may be added provided its unit cost does not exceed one half of the sum of the costs for the three with the lowest, and so forth. Again, in the special case $c_i = c$, nothing is constrained.

The Cournot Iterative Map

It is now time to return to the general case, and consider the dynamics of Cournot action. To this end it is necessary to state how expectations for the residual supplies Q_i are formed in an evolving system in order that each competitor be able to calculate the proper response according to $q_i = \phi_i(Q_i)$ as stated in (2.6) end exemplified in (2.11) and (2.19). Somehow real competitors and modellers alike have to form such expectations, and, as already stated, the assumption closest at hand is that each competitor assumes the others to maintain their previous moves, even if the assumption constantly shows up as being wrong.

This links future to past, so one can write

$$q_i(t+1) = \phi_i(Q_i(t)) \tag{2.46}$$

with time period identifications for the variables. Of course

$$Q(t) = \sum_{i=1}^{i=n} q_i(t), \quad Q_i(t) = Q(t) - q_i(t), \tag{2.47}$$

and likewise for time period $t+1$. These rules indicate how all the quantity variables, individual supply, total supply, and residual supply for each following period are obtained from those of the current. Hence the whole following orbit of the system could in principle be calculated.

For any such system only four possible orbit types which are approached asymptotically after a transient exist. Fixed points (or equilibria), periodic orbits (which always return to the same state after a finite number of steps), quasiperiodic orbits (those that have an infinite period, coming close to previous states periodically but never quite so in finite time), and chaotic orbits (such that are unpredictable, as nearby orbits are separated at an exponential rate, though bent back and kept in a finite area of the phase space).

All these are possible attractors. They can be lonely, or they can coexist. A fixed point can coexist with another fixed point, as demonstrated by Palander (1936, 1939) and Wald (1936), with a periodic orbit (as also shown by Palander), or even with a chaotic attractor. In the case of coexistence, each attractor has its proper basin of attraction, and the separating basin boundaries divide the total phase space in such attraction basins. Basins and basin boundaries can be simple, simply connected areas and curves of finite length, but they can also, even with very simple systems, be very complicated, so called fractals.

These potential attractors, fixed points, periodic, quasiperiodic, and chaotic, can also be unstable, i.e., repellors. They yet remain interesting for describing the dynamics, as they exist and repel all trajectories that come near.

Further, attractors can turn into repellors due to some slight change of a parameter of the model, in which case one speaks of a bifurcation. Modern theory for dynamical systems has got very far in studying and classifying types of bifurcations, as well as of attractors and basins.

Formation of Expectations

In formulating the iterative dynamic system, or map as is the mathematical term, it was assumed that the competitors just assumed all others to retain their previous moves. This is a quite problematic case in an evolving system, except when it is in a stable fixed point.

Observing Chaotic Orbits

So, what can the competitors learn from the actually observed orbit? A chaotic orbit is by definition unpredictable, though deterministic. This is due to the magnification of computational rounding off errors. Determinism just means that the future of the system is totally determined by the initial conditions; but these initial conditions must be known exactly, numerical values with an infinite number of decimals. Such precision is never possible in reality. Now, a slight measurement error can have different consequences. In a system such as focused in traditional dynamics, it just leads to a slight displacement of the orbit. In a chaotic system, orbits starting from different initial conditions, no matter how slight the difference is, need no more than a few moves to end up in completely different parts of the phase space. This is what unpredictability means. For the present context it is obvious that orbits that are never predictable are useless for observation and use in forming expectations. Unfortunately, modern dynamics has shown that this phenomenon tends to be present in even the simplest nonlinear dynamic systems.

Observing Periodic Orbits

Starting at the simplest end, disregarding the obvious case of a stable fixed point, suppose the system actually produces a periodic orbit, such as may be observed and learned by the competitors. Would it then be possible to adapt to this, and produce an orbit of exactly that periodicity? Such an idea is close to the favourite economics idea of rational expectations.

As will now be shown this is always impossible, because such learning and adapting to any periodicity is bound to produce a periodicity *different* from the one learned, unless the orbit is a fixed point. See Puu (2008).

So, suppose there exists a periodic orbit of period T as solution to the dynamic Cournot system. Then, for the factual orbit,

$$q_i(t+T) = q_i(t), \tag{2.48}$$

holds for all time periods t. As the periodicity is assumed to be learned and adjusted to by the competitors, they will form the expectations correctly and respond to observed facts not one period back, but T periods back. Hence,

$$q_i(t+T) = \phi_i(Q_i(t)). \tag{2.49}$$

But, from equations (2.48) and (2.49), then

$$q_i(t) = \phi_i(Q_i(t)), \tag{2.50}$$

which relates the variables in the same time period, and is hence a definition for the Cournot equilibrium fixed point. This relation holds, as assumed, for all time periods t, which shows that the assumed periodic orbit is a fixed point. A fixed point, of course, is a periodic point – of all periodicities. But this simple proof shows that a periodic point which is not a fixed point cannot be learned, adapted to, and yet result in a periodic orbit of the assumed period, unless it is a fixed point.

For the reader who is very fond of the idea of rational expectations, we can add a different argument. Suppose the dynamic system produces an orbit of, say, period 3. This means that the system eventually visits three points here called ***A***, ***B***, ***C*** over and over. Now, suppose the competitors learn this, and adapt. As the system is deterministic, and the same as before, the outcome would be the same sequence, though with two undetermined points in intervening periods. Assuming one again starts in ***A***, the system produces

$$\boldsymbol{A}\,\_\,\_\,\boldsymbol{B}\,\_\,\_\,\boldsymbol{C}\,\_\,\_\,.$$

Now, try to make this the original 3-period orbit. In that case ***B*** should follow the first entry ***A***, and fill out the first blank, but then the third and fifth blanks will be filled out by ***C*** and ***A*** in that order, producing

$$A\,\boldsymbol{B}\,\_\,B\,\boldsymbol{C}\,\_\,C\,\boldsymbol{A}\,\_\,.$$

Finally, the remaining blanks need to be filled out, by ***C***, ***A***, ***B***, producing

$$A\,B\,\boldsymbol{C}\,B\,C\,\boldsymbol{A}\,CA\,\boldsymbol{B}\,.$$

This starts quite correctly with ***A***, ***B***, ***C***, but then comes ***B***, ***C***, ***A***, and finally, ***C***, ***A***, ***B***. It is thus impossible to recognize the sequence as anything but a 9-period orbit. After the last entry the sequence starts all over again.

This shows that if the competitors learn the 3-periodicity they produced before, then they, in fact, produce a 9-period orbit, which, if learned and adapted to, in its turn produces another periodicity. If all the competitors do not learn the periodicity simultaneously, but some are slower in learning and adapting than others, then the result becomes even much more complicated.

The conclusion is that learning and adapting to an observed periodicity by the agents is always bound to change this very periodicity. In a loose way it is similar to the impossibility principle in physics, where the very act of observing and measuring positions and momenta of particles changes the facts to be observed.

It is self-evident, that the change of periodicity does not occur when the periodic orbit is a fixed point with just one point of phase space ***A*** visited over and over.

As a general conclusion; some orbits, such as the chaotic, are too complicated to learn and adapt to; some, such as the periodic, are simple to observe, learn, and adapt to, but when the agents do this, they inevitably change the outcome. To adapt to an observed orbit and make the system produce that very orbit seems to be impossible, unless the orbit is a fixed point.

What then can the competitors do to escape from expecting the others to retain their previous moves, sometimes called naive expectations, which always shows up to be wrong? They can observe trends, increases or decreases from two subsequent period observations, or even curvatures in an attempt to identify approaching turning points. However, analysing such systems is bound to become extremely complex and messy as it multiplies up the order of the system.

In the sequel this simple assumption concerning the formation of expectations will be kept, except when discussing conservative moves in terms of adaptive systems, or Stackelberg leadership, which represents another type of learning, not of the actual orbit, but of the reaction functions of the competitors.

Stability

Cournot equilibria, fixed points of the Cournot iterative map, can, as stated above, be stable or unstable. If they are unstable, they can yet be of interest in characterizing the dynamic system, but they are no attractors.

For establishing the stability of a Cournot equilibrium point of the general iterative system,

$$q_i(t+1) = \phi_i(Q_i(t)), \tag{2.51}$$

one first needs to calculate the derivatives of the reaction functions,

$$\frac{dq_i(t+1)}{dQ_i(t)} = \phi_i' . \tag{2.52}$$

Then the Jacobian matrix reads,

$$J = \begin{bmatrix} 0 & \phi_1' & \cdots & \phi_1' \\ \phi_2' & 0 & \cdots & \phi_2' \\ \vdots & \vdots & \ddots & \vdots \\ \phi_n' & \phi_n' & \cdots & 0 \end{bmatrix}. \tag{2.53}$$

Obviously, as the supplies of all the competing firms enter in the sum defined as residual supply, the same entries appear as off-diagonal elements in each row. As the supply of each firm itself is absent in this definition of residual supply, the diagonal elements are zero.

The next move would be to form the matrix

$$J - \lambda I = \begin{bmatrix} -\lambda & \phi_1' & \cdots & \phi_1' \\ \phi_2' & -\lambda & \cdots & \phi_2' \\ \vdots & \vdots & \ddots & \vdots \\ \phi_n' & \phi_n' & \cdots & -\lambda \end{bmatrix}, \tag{2.54}$$

where I denotes the n by n identity matrix, and the characteristic equation,

$$|J - \lambda I| = 0. \tag{2.55}$$

This nth degree polynomial equation in λ determines the eigenvalues, real or complex, and for stability one must ensure that all these $\lambda_1, \lambda_2, \ldots \lambda_n$ are in the unit circle in the complex plane.

This is a formidable programme, and given the general equations for the Cournot system,

$$f(q_i(t+1) + Q_i(t)) + f'(q_i(t+1) + Q_i(t))q_i(t+1) = C_i'(q_i(t+1)), \tag{2.56}$$

implicit differentiation results in

$$\phi_i' = -\frac{f' + \phi_i f''}{2f' + \phi_i f'' - C_i''}, \tag{2.57}$$

which seems to be exceedingly messy. One would have to calculate the Cournot point coordinates from (2.23), substitute those in the derivatives of the reaction functions (2.57), then those in the Jacobian (2.53), and finally solve the polynomial characteristic equation (2.55). Several of these steps are just impossible in view of the general procedures of mathematical economics.

That much just to determine the stability of a single fixed point! How could one at all deal with the orbits resulting from an unspecified iterative map with just a few qualitative properties? These facts emphasize the absolute need to work with specified global models resulting in credible closed form reaction functions.

Example 1: Linear Demand

It is therefore good to return to the two examples. In the literature Theocharis (1959) is generally still credited for having been the first to show that a Cournot equilibrium in an oligopoly with a linear demand function and constant unit costs for the competitors is destabilised when their number exceeds three, whereas stability becomes neutral with three competitors. Then the (unique) fixed pint is not yet unstable, but any slight deviation from it puts up an endless oscillatory motion; each such motion being neutrally stable as well.

Palander (1939), whose more important contributions will be dealt with in a later chapter, stated exactly the same 20 years earlier. Unfortunately, the article was written in Swedish, a language even present day Anglo-Saxon oriented Swedish economists neither read nor cite.

The stability issue, which was extremely complicated in the general case, is now equally simple. From the reaction functions,

$$q_i(t+1) = \phi_i(Q_i(t)) = \frac{a - c_i}{2b} - \frac{1}{2} Q_i(t) . \tag{2.58}$$

The derivatives are always

$$\frac{dq_i(t+1)}{dQ_i(t)} = \phi_i' = -\frac{1}{2} , \tag{2.59}$$

i.e., constant, so that one does not even need the particular coordinates of the Cournot equilibrium point to put up the n by n Jacobian matrix

$$J = \begin{bmatrix} 0 & -\frac{1}{2} & \cdots & -\frac{1}{2} \\ -\frac{1}{2} & 0 & \cdots & -\frac{1}{2} \\ \vdots & \vdots & \ddots & \vdots \\ -\frac{1}{2} & -\frac{1}{2} & \cdots & 0 \end{bmatrix}. \tag{2.60}$$

The characteristic equation is easily obtained as an nth degree polynomial that factorizes into

$$|J - \lambda I| = P^n(\lambda) = (\lambda - \tfrac{1}{2})^{n-1}(\lambda + \tfrac{1}{2}(n-1)). \tag{2.61}$$

The eigenvalues are

$$\lambda_1, \ldots \lambda_{n-1} = \tfrac{1}{2}, \lambda_n = -\tfrac{1}{2}(n-1). \tag{2.62}$$

The first $n-1$ are equal to $\frac{1}{2}$ and are well within the unit circle of the complex plane, but the last one $\frac{1}{2}(n-1)$ falls outside it when $n > 3$. For $n = 3$, $\lambda_n = -1$, so this is a boundary case.

The multiple eigenvalues are associated with differences of the variables. They always indicate stability, even when the system diverges. This means that, in any motion, differences between the behaviour of the firms are evened out, so that the actions are asymptotically coordinated. The last eigenvalue is associated with total market supply, and it is it that brings instability to the system when the number of competitors grows.

Theocharis also gave the complete global solution to the system, but there is no need to present it here, because an unstable linear system always just explodes, resulting in infinite deviations, negative supplies, and negative prices. His solution is therefore not factually relevant. The global dynamics of the piecewise linear case, where account is taken of nonnegativity of price, supplies, and profits, was only analysed a few years ago. See Canovas et al. (2008).

Palander too, was only concerned about the instability, but he stated the results completely, as mentioned, 20 years before Theocharis. In 1939 he wrote "*as a condition for equilibrium with a certain number of competitors to be stable to exogenous disturbances, one can stipulate that the derivative of the reaction function* ϕ' *must be such that the condition* $|(n-1)\phi' < 1|$ *holds. If this criterion is applied to, for instance, the case with a linear demand function and constant marginal costs, the equilibria become unstable as soon as the number of competitors exceeds three. Not even in the case of three competitors will equilibrium be restored, rather there remains an endless oscillation*".

As for the global dynamics of the piecewise linear case, not much exciting occurs. There are just the possibility of a stable monopoly, or a duopoly; otherwise two-period oscillations where all the firms move in phase are the attractors. If the number of competitors exceeds three, all firms drop out every second period. This might seem to present an excellent case for learning the orbit by some competitors who might try to move out of phase, even attempting to become monopolists every second period. But then one encounters the problem stated above that this learning and adaptation alters the periodicity itself.

Which firms will stay on the market depends, as was shown in the section on Cournot equilibria, on the cost structure. It may occur that, even with very numerous firms, the cost structure among them might be such that only one or two firms with cost advantages stay in the Cournot equilibrium; all the other being bound to drop out in the dynamic process.

Example 2: Isoelastic Demand

From (2.35) the main branches of the reaction functions in the case of isoelastic demand were

$$q_i(t+1) = \phi_i(Q_i(t)) = \sqrt{\frac{Q_i(t)}{c_i}} - Q_i(t), \tag{2.63}$$

so the derivatives,

$$\frac{dq_i(t+1)}{dQ_i(t)} = \phi_i' = \frac{1}{2}\frac{1}{\sqrt{c_i Q_i(t)}} - 1 \tag{2.64}$$

are easily obtained. Further, from (2.37), in Cournot equilibrium,

$$Q_i = c_i Q^2 . \tag{2.65}$$

Hence,

$$\phi_i' = \frac{1}{2}\frac{1}{c_i Q} - 1 \tag{2.66}$$

in the Cournot fixed point. Further, it was true that

$$Q = \frac{n-1}{n}\frac{1}{c}, \tag{2.67}$$

where

$$c = \frac{1}{n}\sum_{i=1}^{i=n} c_i \tag{2.68}$$

was average marginal cost. Using (2.67) for Q,

$$\phi_i' = \frac{1}{2}\frac{n}{n-1}\frac{c}{c_i} - 1 \tag{2.69}$$

in Cournot equilibrium.

Agiza (1998) and Ahmed et al. (1998) wanted to demonstrate that the Cournot equilibrium is destabilised in this case when the number of competitors exceeds four, and is neutrally stable when the number is exactly four. The frontier of destabilisation is hence just pushed one competitor further as compared to the case of a linear demand function. Their proof could not be completed without changing the demand function to

$$p = \frac{1}{Q} + p_0 , \tag{2.70}$$

through adding a constant. This, however, unlike the original case of the isoelastic function, is not derived from any basic microeconomics of consumers maximizing their utility functions. Further, note that this change has no connection with the problem discussed that the isoelastic demand function is unsuitable for dealing with the case of monopoly; to avoid problems in that connection, one would have to add a constant to supply, rather than to price. But, unfortunately, this is not based on the theory of the consumer either.

However, there is a simple way to arrive at the conclusions of Agiza (1998) and Ahmed et al. (1998) with the original demand function. That is to assume that the competitors are identical, so that all $c_i = c$.

Then

$$\phi_i' = -\frac{1}{2}\frac{n-2}{n-1}, \tag{2.71}$$

which facilitates things a lot. The n by n Jacobian matrix for the Cournot equilibrium point now becomes

$$J = \begin{bmatrix} 0 & -\frac{1}{2}\frac{n-2}{n-1} & \cdots & -\frac{1}{2}\frac{n-2}{n-1} \\ -\frac{1}{2}\frac{n-2}{n-1} & 0 & \cdots & -\frac{1}{2}\frac{n-2}{n-1} \\ \vdots & \vdots & \ddots & \vdots \\ -\frac{1}{2}\frac{n-2}{n-1} & -\frac{1}{2}\frac{n-2}{n-1} & \cdots & 0 \end{bmatrix}, \tag{2.72}$$

and the characteristic equation

$$|J - \lambda I| = P^n(\lambda) = \left(\lambda - \frac{1}{2}\frac{n-2}{n-1}\right)^{n-1}\left(\lambda + \frac{1}{2}(n-2)\right) = 0. \tag{2.73}$$

Again, the polynomial factorizes, and among the eigenvalues,

$$\lambda_1, \ldots \lambda_{n-1} = \frac{1}{2}\frac{n-2}{n-1}, \lambda_n = -\frac{1}{2}(n-2), \tag{2.74}$$

the first, of multiplicity $n - 1$, are inside the unit circle in the complex plane, whereas the last one comes outside for $n > 4$, and is hence associated with instability.

The present model with isoelastic demand offers better perspectives for interesting dynamics in global analysis than the case with linear demand, but, quite as in the linear case, the Cournot equilibrium is destabilsed when the number of competitors exceeds a very small number, though now four in stead of three.

The significance of the result, attributed to Theocharis (1959), has relevance for the issue of increasing competition transforming the market from monopoly over oligopoly to perfect competition. This means that with an increasing number of competitors, the Cournot equilibrium approaches marginal cost pricing and elimination of profits for the marginal firm (the one with highest unit cost). But, this, which, as was shown above, happens in both models, is of little interest if in this same process equilibrium is destabilised. If the system is no longer attracted to the Cournot equilibrium it does not matter that it transforms into a perfect competition equilibrium.

Yet, intuitively, one would like to keep the possibility of oligopoly seamlessly transforming into perfect competition. The question, which seems, not to have been addressed is; why does this happen with different reasonable demand functions?

The clue lies in the assumptions on the cost side, constant marginal and average variable costs. These emerge from production under constant returns. But a firm producing under constant returns has in principle infinite capacity. In perfect competition, a market price, taken as constant by the individual firms, which exceeds a constant unit cost by the tiniest fraction, makes it possible to blow up the profit to any value just through multiplying up the scale of operation.

Admitting this, one may say that destabilisation due to adding more and more infinite sized firms is not very surprising; nor is it relevant to the issue of transforming oligopoly into perfect competition. It has been implicit that the comparison should be between cases of few large firms versus many small firms. But large and small cannot be modelled with constant unit costs. One would need variable, not constant, returns, preferably including capacity limits for such modelling.

This can, as shown in a following chapter, be modelled in different ways. In the piecewise linear model, one could keep constant unit costs, but at a blunt capacity limit let the cost jump up to infinity. It is also possible to use the CES production function, with capital fixed through an act of investment, to obtain a cost function that asymptotically goes to infinity at a capacity limit due to the fixed capital. This makes it necessary to include assumptions of capital formation, and paves the ground for endogenous modelling of how capital formation and competition may evolve in a branch. See Puu and Panchuk (2009).

References

Agiza, H. N. (1998). Explicit stability zones for Cournot games with 3 and 4 competitors. *Chaos, Solitons & Fractals, 9*, 1955–1966.

Ahmed, E., & Agiza H. N. (1998). Dynamics of a Cournot game with *n* competitors. *Chaos, Solitons & Fractals, 9*, 1513–1517.

Canovas, J. S., Puu, T., & Ruiz, M. (2009). The Cournot–Theocharis problem reconsidered. *Chaos, Solitons & Fractals, 37*, 1025–1039.

Cournot, A. (1838). *Récherces sur les principes mathématiques de la théorie des richesses.* Paris: Dunot.

Palander, T. F. (1936). Instability in competition between two sellers, Abstracts of papers presented at the research conference on economics and statistics held by the Cowles Commission at Colorado College, *Colorado College Publications*, General Series No. 208, Studies Series No. 21.

Palander, T. F. (1939). Konkurrens och marknadsjämvikt vid duopol och oligopol. *Ekonomisk Tidskrift, 41*, 124–145, 222–250.

Puu, T. (1991, 2004). Chaos in duopoly pricing. *Chaos, Solitons & Fractals, 1*, 573–581, republished in: J. B. Rosser (Ed.), Complexity in Economics: *The International Library of Critical Writings in Economics,* Vol. 174. Cheltenhan Edward Elgar.

Puu, T. (2008). Rational expectations and the Cournot-Theocharis problem, *Discrete Dynamics in Nature and Society* ID, 32103, 1–11.

Puu, T., & Panchuk, A. (2009). Oligopoly and stability. *Chaos, Solitons, & Fractals, 41*, 2505–2516.

Robinson, J. (1933). *The Economics of Imperfect Competition.* London: Macmillan.

Theocharis, R. D. (1959). On the stability of the Cournot solution on the oligopoly problem. *Review of Economic Studies, 27*, 133–134.

Tramontana, F., Gardini, L., & Puu, T. (2010). New properties of the Cournot duopoly with isoelastic demand and constant unit costs. Chaos, Solitons, & Fractals (in press). (submitted).

Wald, A. (1936). Über einige Gleichungssysteme der mathematischen Ökonomie. *Zeitschrift für Nationalökonomie, 7*, 637–670.

3 Duopoly and Complex Dynamics

Introduction

As mentioned in Chap. 1, Rand (1978) conjectured that with suitable shapes of the reaction functions the outcome of dynamic Cournot (1838) duopoly would be chaotic. His purely mathematical treatment does not, however, include any substantial economic assumptions under which this becomes true. In what follows a very simple duopoly model, already encountered in Chap. 2, based on traditional microeconomis, will be discussed.

It is the case with isoelastic demand and constant marginal costs. The other case, with linear demand, presents no interesting scenarios. It is destabilised with an increasing number of competing firms, but the global scenario is always a two-period oscillating orbit. The isoelastic case, on the contrary, produces a period-doubling cascade of bifurcations, ending in chaos. See Puu (1991, 2004). This has nothing to do with the discussed destabilisation due to addition of competitors; this happens in duopoly. The facts that trigger the sequence of events are increasing inequality of the competitors' unit costs.

After discussing the basic model in some detail, an adaptive variant of it will be discussed. The competitors then do not immediately jump to their calculated best replies, but only move towards these replies from their previous positions, knowing that their expectations never come true which calls for some conservatism.

It can be emphasized that the adaptive variant removes the anomaly present in the assumption of isoelastic demand. As mentioned, it is not suitable for dealing with monopoly or collusion, due to the fact that total market revenue is fixed so that monopolists do best to produce nothing to avoid cost.

The case here is duopoly, but even then, as will be shown, there must be zero branches of the reaction functions. Under certain circumstances, the iterative process may end up at the collusive state (the origin), or at a two-period

T. Puu, *Oligopoly: Old Ends – New Means*, DOI 10.1007/978-3-642-15964-0_3,

oscillating collusive state where the firms alternate in dropping out and leaving the competitor as monopolist (alternating monopoly). The curious fact is that, even if the intersection point of reaction functions at the origin is totally unstable, with both reaction functions having infinite slope, they are yet weak attractors with their proper basins of attraction. Any tiny perturbation from such an equilibrium, or an assumption that the firms never go down to zero in production but retain some "epsilon" stand by output, do indeed fix the problem, but then the exact numerical value of such "epsilons" or perturbations influence the dynamic of the process.

The adaptive hypothesis clears away this problem as the competitors never move down to the zero branches. In addition, it makes the system truly two-dimensional, whereas in the original model a composition of two iterations results in separate processes for each variable alone. The phenomena in phase space then are just combinations of these independent processes in each variable alone.

The Cournot Model

Assume an isoelastic demand function, such that price, denoted p, is reciprocal to the total supply, made up of the supplies of the two competitors, denoted $q_1 = Q_2$ and $q_2 = Q_1$,

$$p = \frac{1}{q_1 + q_2}. \tag{3.1}$$

Next, suppose the duopolists produce with constant marginal costs, denoted c_1 and c_2 respectively. The profits of the two firms become,

$$\Pi_1(q_1, q_2) = \frac{q_1}{q_1 + q_2} - c_1 q_1, \tag{3.2}$$

$$\Pi_2(q_1, q_2) = \frac{q_2}{q_1 + q_2} - c_2 q_2. \tag{3.3}$$

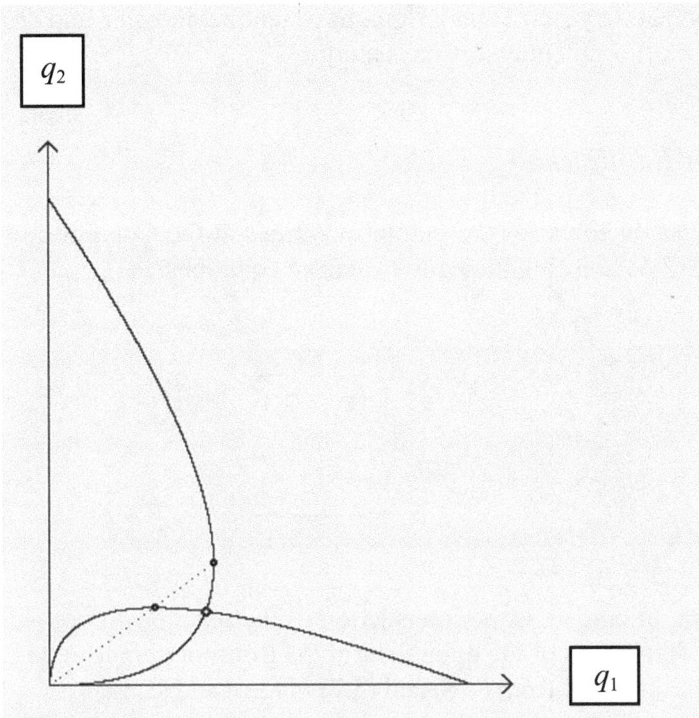

Fig. 3.1 Reaction curves and the Cournot point

The first firm maximizes $\Pi_1(q_1,q_2)$ with respect to q_1, the second $\Pi_2(q_1,q_2)$ with respect to q_2. Equating the partial derivatives to zero, one can solve for the reaction functions,

$$q_1 = \phi_1(q_2) = \sqrt{\frac{q_2}{c_1}} - q_2, \tag{3.4}$$

$$q_2 = \phi_2(q_1) = \sqrt{\frac{q_1}{c_2}} - q_1. \tag{3.5}$$

A check of the second order conditions, in fact, testifies that we always deal with local profit maxima, provided quantities are positive, as indeed they should be. The reaction functions are displayed in Fig. 3.1. Their general outline is that they start at the origin, have unique maxima, and drop to zero again. The off origin intersection is unique.

Cournot Equilibrium

One can easily solve for the output quantities at the Cournot point, taking (3.4) and (3.5) as a simultaneous system of equations:

$$q_1 = \frac{c_2}{(c_1 + c_2)^2} \tag{3.6}$$

$$q_2 = \frac{c_1}{(c_1 + c_2)^2} \tag{3.7}$$

This point, of course, is the intersection of the reaction curves as shown in Fig. 3.1. The profits of the duopolists at the Cournot point can be calculated by substituting back from (3.6) and (3.7) in (3.2) and (3.3):

$$\Pi_1 = \frac{c_2^2}{(c_1 + c_2)^2} \tag{3.8}$$

$$\Pi_2 = \frac{c_1^2}{(c_1 + c_2)^2} \tag{3.9}$$

The Iterative Process

As mentioned in the introduction, Cournot also designed an iterative dynamic process. Assuming the simple expectation formation projecting past competitive moves into the future, one needs to lag the variables. So, (3.4) and (3.5) must be written:

$$q_1(t+1) = \phi_1(q_2(t)) = \sqrt{\frac{q_2(t)}{c_1}} - q_2(t) \tag{3.10}$$

$$q_2(t+1) = \phi_2(q_1(t)) = \sqrt{\frac{q_1(t)}{c_2}} - q_1(t) \tag{3.11}$$

This pair of equations is the central piece of the iterative process we are going to study. However, we have to deal with two special questions:

First, one has to recognize that there is a problem with (3.10) and (3.11) as stated when $c_1 q_2(t) > 1$ or $c_2 q_1(t) > 1$, as then $q_1(t+1) < 0$ or $q_2(t+1) < 0$ would be returned. Mathematically such *negative outputs* even result in positive maximum profits, as negative costs dominate over negative revenues, but in terms of economics this is of course *nonsense*, and has to be ruled out. Further, any negative output would in the next iteration return a complex number due to the square root. So we rephrase (3.10) and (3.11) as

$$q_1(t+1) = \begin{cases} \phi_1(q_2(t)) & c_1 q_2(t) \le 1 \\ 0 & c_1 q_2(t) > 1 \end{cases} \tag{3.12}$$

$$q_2(t+1) = \begin{cases} \phi_2(q_1(t)) & c_2 q_1(t) \le 1 \\ 0 & c_2 q_1(t) > 1 \end{cases} \tag{3.13}$$

It is a matter of the parameters c_1 and c_2 whether the entire orbit can be calculated, using the regular branches of the reaction functions (3.10) and (3.11) only, or whether the zero branches of (3.12) and (3.13) have to be included as well. To be precise, it is a matter of the marginal cost *ratio* c_2 / c_1 alone. As we will find later, many other properties too depend on this ratio alone, and, as a matter of fact, the whole process could have been cast in terms of one ratio parameter only. But for the sake of symmetry we keep both.

The maximum points for $\phi_1(q_2)$, $\phi_2(q_1)$, can be easily calculated from (3.4) and (3.5) through taking the derivatives, equating to zero, and solving. Substituting back in (3.4) and (3.5), we then easily obtain the respective maximum values

$$\phi_1\left(\frac{1}{4c_1}\right)=\frac{1}{4c_1}, \tag{3.14}$$

$$\phi_2\left(\frac{1}{4c_2}\right)=\frac{1}{4c_2}. \tag{3.15}$$

Obviously, the maxima are located on the diagonal in phase space as shown in Fig. 3.1.

Whether the process visits the zero branches or not is a matter of whether these maximum values exceed the values at which the other reaction function becomes zero, i.e.,

$$\phi_2(q_1)=0 \Rightarrow q_1=\frac{1}{c_2}, \tag{3.16}$$

$$\phi_1(q_2)=0 \Rightarrow q_2=\frac{1}{c_1}. \tag{3.17}$$

Hence, if and only if,

$$\frac{c_2}{c_1}<4, \quad \frac{c_1}{c_2}<4 \tag{3.18}$$

hold, are the zero branches never visited. Once a zero branch is visited, there arise additional attractors besides those described in the sequel, which will, however not be discussed here. See Tramontana et al. (2010).

There is also a second issue which we have to settle. In earlier literature great emphasis was put on whether duopolists react *simultaneously* or *sequentially* according the reaction functions such as (3.10) and (3.11). In order to deal with this, first note that, substituting (3.11) in (3.10), or substituting (3.10) in (3.11), one obtains

$$q_1(t+2) = \phi_1(\phi_2(q_1(t))) = \phi_1 \circ \phi_2(q_1(t)), \tag{3.19}$$

$$q_2(t+2) = \phi_2(\phi_1(q_2(t))) = \phi_2 \circ \phi_1(q_2(t)). \tag{3.20}$$

These are independent iterations, each in one phase variable alone, though based on a period *twice as long* as the fundamental period introduced from the outset. Given initial phase variable values *for two subsequent periods*, one can compute the complete trajectory for each variable.

Obviously the functions $\phi_1 \circ \phi_2$ and $\phi_2 \circ \phi_1$ are autonomous iterations, so one gets the same trajectories whether the firms adjust simultaneously or in sequence. The difference between simultaneous and sequential adjustment hence only lies in the way of pairing the two iteration series together. There also is a difference in terms of the freedom to choose initial conditions. If the adjustment is sequential, then the second iterate can be calculated from only one initial condition, and one *cannot* choose all initial points in the phase space, just points located on the graphs of the reaction functions. This being said, it should be noted that the system may itself settle on successive adjustment.

Stability of the Cournot Point

The only fixed point of the iteration (3.10) and (3.11), except the origin, is, of course, the Cournot equilibrium point (3.6) and (3.7). To find out something about the stability of this point, calculate the derivatives of the functions (3.4) and (3.5):

$$\phi_1'(q_2(t)) = \frac{1}{2}\frac{1}{\sqrt{c_1 q_2(t)}} - 1 \tag{3.21}$$

$$\phi_2'(q_1(t)) = \frac{1}{2}\frac{1}{\sqrt{c_2 q_1(t)}} - 1 \tag{3.22}$$

At the Cournot point with coordinates (3.6) and (3.7), expressions (3.21) and (3.22) simplify to:

$$\phi_1'(q_2) = \frac{c_2 - c_1}{2c_1} \tag{3.23}$$

$$\phi_2'(q_1) = \frac{c_1 - c_2}{2c_2} \tag{3.24}$$

Loss of stability of the Cournot point occurs when:

$$\left|\phi_1'(q_2)\phi_2'(q_1)\right| = \left|\frac{(c_2 - c_1)}{2c_1}\frac{(c_1 - c_2)}{2c_2}\right| = 1 \tag{3.25}$$

Expanding (3.25), we get:

$$(c_1 - c_2)^2 = 4c_1c_2 \tag{3.26}$$

Observe that this condition can be solved for the ratio:

$$\frac{c_1}{c_2} \quad or \quad \frac{c_2}{c_1} \quad = \quad 3 \pm 2\sqrt{2} \tag{3.27}$$

Thus, whenever one of the ratios of the marginal costs of the duopolists falls outside the interval bounded by the two values specified in (3.27), the Cournot point will not be stable. The two roots happen to be reciprocal, so there is nothing odd in stating this condition for *any* of the ratios. As we already noted other critical conditions as well depend on the ratio of marginal costs only, so there is just one single free control parameter.

Periodic Points and Chaos: Numerical study

As this control parameter passes either of the critical values specified in (3.27), the fixed point is replaced by a two period cycle. This should be so understood that q_1 and q_2 each oscillate between two values. The two variables themselves, of course, take on different pairs of values in the oscillatory process.

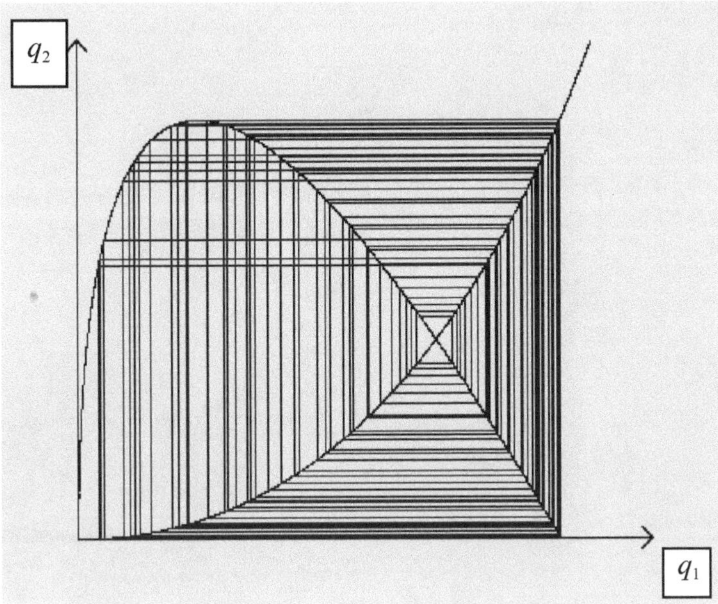

Fig. 3.2 Chaotic Cournot duopoly adjustments

After the first cycle appears, there is a period doubling cascade to chaos. In Fig. 3.2 the cobweb of a chaotic process of adjustments is displayed in the same type of diagram as Fig. 3.1.

Figure 3.3 we illustrates the bifurcation diagram. Interesting things begin to happen when the ratio c_2 / c_1 exceeds the critical value $3+2\sqrt{2} \approx 5.8284$ indicated in (3.27), or falls short of its reciprocal. Until then there is just a stable fixed point. The smooth functions (3.10) and (3.11) alone are relevant for the scenario illustrated.

As mentioned, there also exist degenerate attractors whenever c_2 / c_1 exceeds the critical case of 4, but those will not be discussed here. There is a further critical value when the cost ratio exceeds 6.25 or falls short of its reciprocal, when the chaotic attractor is turned into a repellor. See Tramontana et al. (2010). For this reason the bifurcation diagram for the marginal cost ratio is cut off at 6.25.

In the range shown, the fixed point undergoes a period doubling bifurcation cascade to chaos. Depicted is the asymptotic behaviour of both the q_1 variable (brighter shade) and the q_2 variable (darker shade), against the marginal cost ratio c_2 / c_1.

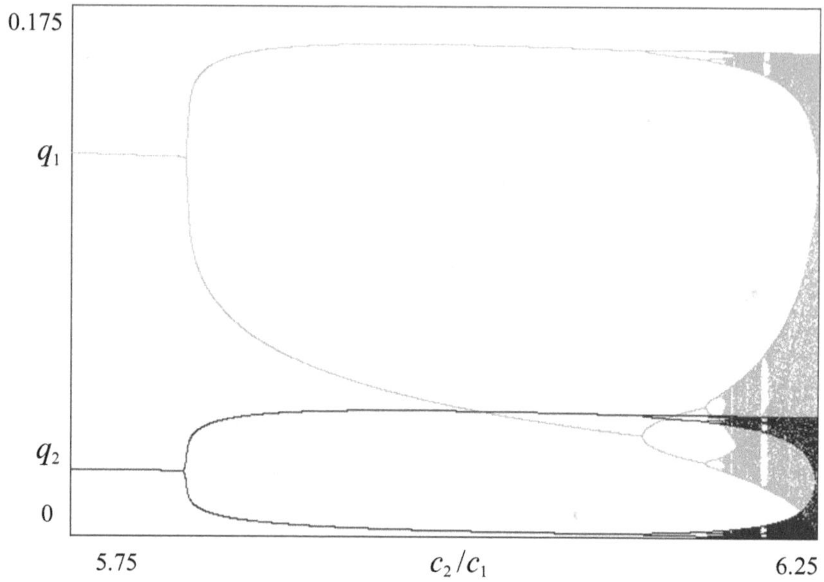

Fig. 3.3 Bifurcation diagram. Amplitudes plotted vs. marginal cost ratio

As already noted, the behaviour of the system is symmetric as to this ratio and its reciprocal c_2 / c_1. So shifting to the latter ratio would give the same bifurcation diagram. Only the roles of the phase variable amplitudes are swapped. Further, for the same reason, we need not extend the parameter ratio axis to the reciprocal of 6.25, because its leftmost part would only be a compressed mirror image of the information already shown.

The general appearance of the bifurcation diagram is very like that of the extensively studied logistic map. There is the period doubling cascade to chaos, as well as the windows of order in the chaotic region.

That much for the independent iterations (3.19) and (3.20). If we study the phenomena for both variables in the q_1, q_2-phase space shown in Fig. 3.1, then they combine to produce coexistent cycles. Figures 3.4 and 3.5 show two such cases.

In Fig. 3.4, the parameter ratio is $c_2 / c_1 = 6.15$, which, as seen in Fig. 3.3, produces 4-period cycles in each variable alone. They obviously combine to produce two different coexisting 8-period cycles in Fig. 3.4.

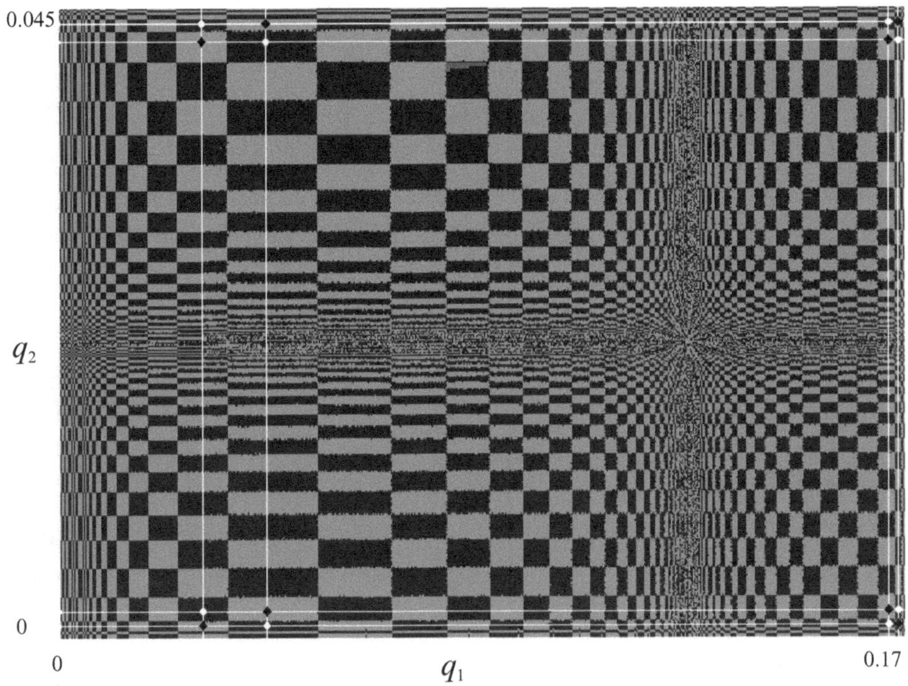

Fig. 3.4 Two coexistent 8-period cycles and their basins of attraction. Parameter value $c_2 / c_1 = 6.15$

The locations of the four independent periodic points in each variable are indicated by the four horizontal and vertical lines, whose intersections produce a 4 by 4 lattice. The periodic points are located on this 16-point lattice. The cycles are indicated by the differently marked dots. Also seen is a chequer-board pattern of fragmented attraction basins corresponding to these coexistent cycles. Notably, the mesh becomes smaller in the neighbourhood of the Cournot equilibrium point. Close to that point infinitesimal differences in initial conditions make the system go for one or another of the coexistent periodic attractors.

The case in Fig. 3.5 is taken for $c_2 / c_1 = 6.212$, which comes from the largest window of order seen in Fig. 3.3. In this window (3.10) and (3.11) produce 6-period cycles for each variable alone. These are again indicated by the six horizontal and vertical lines. Their intersection forms a 6 by 6 lattice. Located on it are three different sets of dots indicated by different characters; in all three coexistent 12-period cycles. Corresponding to the cycles are three attraction basins drawn in different shade. The basins are now even more fragmented as compared to the previous picture.

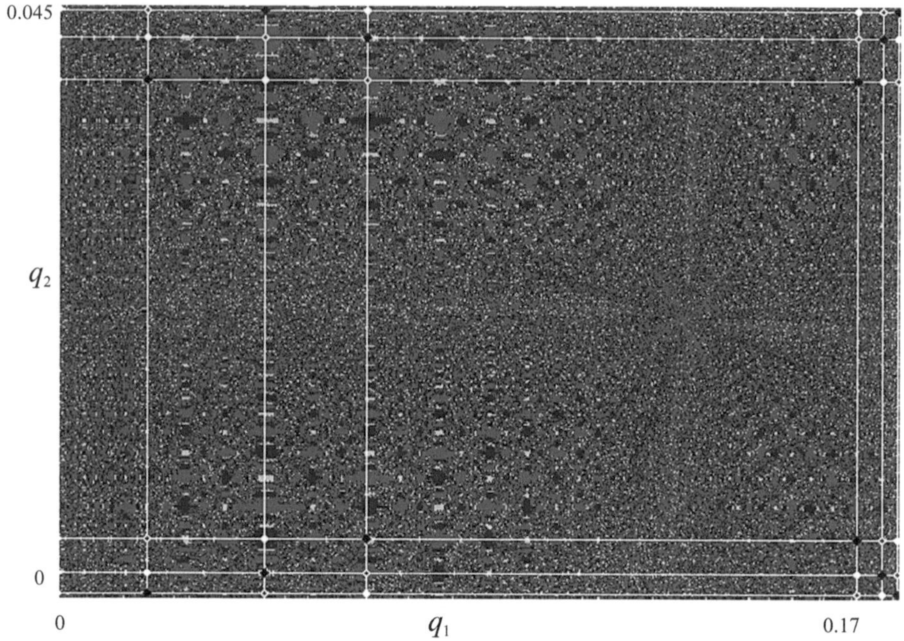

Fig. 3.5 Three coexistent 12-period cycles and their basins of attraction. Parameter value $c_2 / c_1 = 6.212$

In both cases, if there are n-period cycles for each variable alone (n even, as it is in this period doubling cascade), then they combine to produce $n/2$ different coexistent $2n$-period cycles. That this always is the case with maps such as (3.10) and (3.11) has been rigorously proved. See Bischi et al. (2000), where the corresponding results for odd-period n can also be found.

The bifurcation sequence leading to chaos, obviously has been of the period-doubling type. In the sequel we will encounter different types of bifurcations when the Cournot point loses stability. Another typical feature of the (3.10) and (3.11) type of map is the character of the fragmented attraction basins, which entirely consist of *rectangular* areas, getting smaller and smaller in the neighbourhood of the (unstable) Cournot point, so producing a very complex fragmentation, particularly close to this point.

In both Figs. 3.4 and 3.5, the periodic points cluster in four rectangles close to the corners of phase space. When the process is chaotic, these become chaotic rectangles dense of attractor points. This can be seen in Fig. 3.6. Eventually, as the parameter ratio grows towards 6.25, the rectangles grow, and finally merge into one single chaotic rectangle.

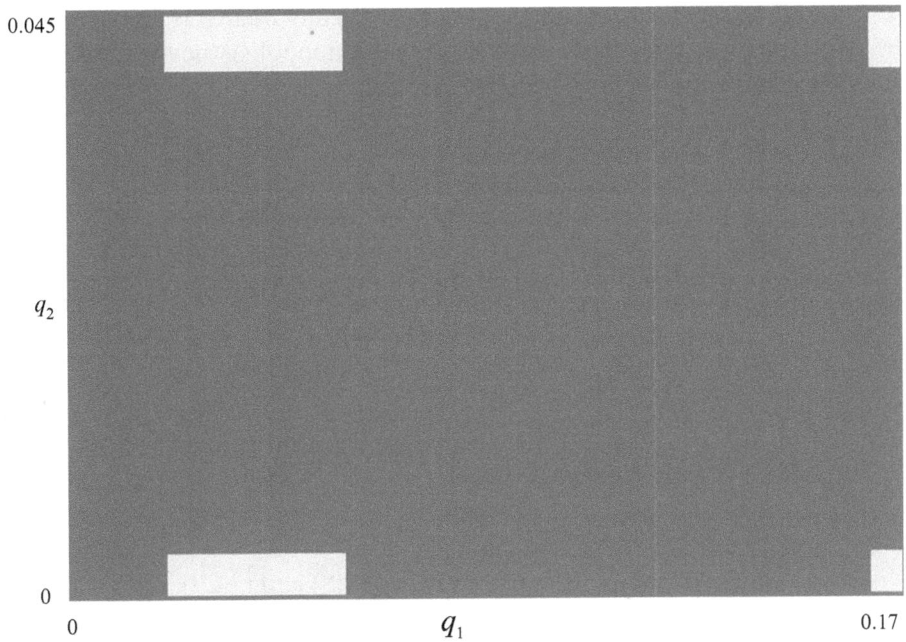

Fig. 3.6 Four piece chaotic attractor. Parameter value $c_2 \,/\, c_1 = 6.2$

Adaptive Expectations

As mentioned, the iterations considered up to now are actually two independent ones. To make the map really two-dimensional, assume that the duopolists do not immediately jump to their new optimal positions, but adjust their previous decisions *in the direction of* the new optimum. As mentioned, this also removes the problem of the zero branch, as the competitors only move *in the directions* of the calculated best replies.

Assume adjustment speeds α and β in the open interval $]0,1[$, and redefine the reaction functions,

$$q_1(t+1) = \phi_1(q_1(t), q_2(t)) = (1-\alpha)q_1(t) + \alpha\left(\sqrt{\frac{q_2(t)}{c_1}} - q_2(t)\right) \quad (3.28)$$

$$q_2(t+1) = \phi_2(q_1(t), q_2(t)) = (1-\beta)q_2(t) + \beta\left(\sqrt{\frac{q_1(t)}{c_2}} - q_1(t)\right) \quad (3.29)$$

With the adjustment speeds unitary, the case already treated is recovered. In the contrary case, with those speeds zero, the duopolists never revise any decision taken. Intermediate values, between zero and unity, produce the interesting cases.

Again the problems arising when the expressions within large parentheses become negative must be dealt with. So, redefine the system,

$$q_1(t+1) = \Phi_1(q_1(t), q_2(t)) = \begin{cases} \phi_1(q_1(t), q_2(t)) & c_1 q_2(t) \le 1 \\ (1-\alpha) q_1(t) & c_1 q_2(t) > 1 \end{cases} \quad (3.30)$$

$$q_2(t+1) = \Phi_2(q_1(t), q_2(t)) = \begin{cases} \phi_2(q_1(t), q_2(t)) & c_2 q_1(t) \le 1 \\ (1-\beta) q_2(t) & c_2 q_1(t) > 1 \end{cases} \quad (3.31)$$

It is easily checked that the adaptive system (3.28) and (3.29), or (3.30) and (3.31), has the same fixed points as the original system without adaptation (3.19) and (3.20), i.e., the always unstable origin, and the Cournot equilibrium point, whose coordinates were given in (3.6) and (3.7). The Cournot point may again be stable, or it may lose stability. The stability of various solutions, such as the Cournot fixed point and the cycles now depend, not only on the marginal cost ratio c_1 / c_2, but on the adjustment speeds α, β as well.

The Neimark Bifurcation

Compared to the previous model, the stability loss now occurs through a Neimark–Sacker bifurcation. One can study the stability around any point in phase space by linearising the system (3.28) and (3.29). The Jacobian matrix for the linearised system then becomes,

$$J(q_1(t), q_2(t)) = \begin{bmatrix} 1-\alpha & \alpha\left(\dfrac{1}{2\sqrt{c_1 q_2(t)}} - 1\right) \\ \beta\left(\dfrac{1}{2\sqrt{c_2 q_1(t)}} - 1\right) & 1-\beta \end{bmatrix} \quad (3.32)$$

In the Cournot point, substituting for the phase variables from (3.6) to (3.7), the Jacobian simplifies to,

$$J = \begin{bmatrix} 1-\alpha & \alpha\dfrac{c_2-c_1}{2c_1} \\ \beta\dfrac{c_1-c_2}{2c_2} & 1-\beta \end{bmatrix}. \tag{3.33}$$

At the threshold of loss of stability for the Cournot point, the determinant of the Jacobian (3.33) is unitary. Putting $|J(q_1, q_2)| = 1$, and simplifying, one gets the condition:

$$(c_1 - c_2)^2 = 4c_1c_2\left(\frac{1}{\alpha} + \frac{1}{\beta} - 1\right) \tag{3.34}$$

Note that, with the adjustment speeds unitary, condition (3.34) is identical to (3.26), as indeed it should. The graph of (3.34) is the bifurcation curve in parameter space.

The characteristic equation is obtained from the Jacobian matrix (3.33) by putting:

$$\begin{vmatrix} 1-\alpha-\lambda & \alpha\dfrac{c_2-c_1}{2c_1} \\ \beta\dfrac{c_1-c_2}{2c_2} & 1-\beta-\lambda \end{vmatrix} = 0 \tag{3.35}$$

where λ denote the eigenvalues. Written out (3.35) becomes:

$$\lambda^2 - (2-\alpha-\beta)\lambda + (1-\alpha)(1-\beta) + \alpha\beta\frac{(c_1-c_2)^2}{4c_1c_2} = 0 \tag{3.36}$$

The coefficient of the linear term is the negative of the trace, the constant term is the determinant of the Jacobian (3.33).

Solving (3.36), the eigenvalues are

$$\lambda_{1,2} = \frac{1}{2}(2-\alpha-\beta) \pm \frac{1}{2}\sqrt{\alpha^2+\beta^2-\alpha\beta\frac{c_1^{\,2}+c_2^{\,2}}{c_1c_2}} \tag{3.37}$$

To see whether they are real or complex on the bifurcation curve (3.34), substitute from (3.34) for the quotient under the root sign in (3.37), obtaining:

$$\lambda_{1,2} = \tfrac{1}{2}(2-\alpha-\beta) \pm \tfrac{1}{2}\sqrt{(\alpha+\beta)(\alpha+\beta-4)} \tag{3.38}$$

As the remaining parameters are adjustment speeds which only take values in the unit interval, i.e. $0 < \alpha, \beta < 1$, the expression under the root sign is negative at the moment of bifurcation. The eigenvalues are hence complex conjugates, and the bifurcation accordingly of the Neimark–Sacker type.

At the moment of bifurcation, the complex eigenvalues cross the unit circle in the complex plane. Hence one can also write (3.38) as

$$\lambda_{1,2} = \cos\theta \pm \mathrm{i}\sin\theta, \tag{3.39}$$

where, as usual, $\mathrm{i} = \sqrt{-1}$, and

$$\cos\theta = \tfrac{1}{2}(2-\alpha-\beta) \tag{3.40}$$

$$\sin\theta = \tfrac{1}{2}\sqrt{(\alpha+\beta)(4-\alpha-\beta)} \tag{3.41}$$

The bifurcation can be to a periodic, or to a nonperiodic orbit, depending on whether θ is a rational, or an irrational multiple of 2π.

The Bifurcation Diagram

Numerical work will indicate that the Neimark bifurcation for the present model is of the *subcritical* type, not the supercritical. In a supercritical bifurcation from a fixed point, the point loses stability and is replaced by nearby

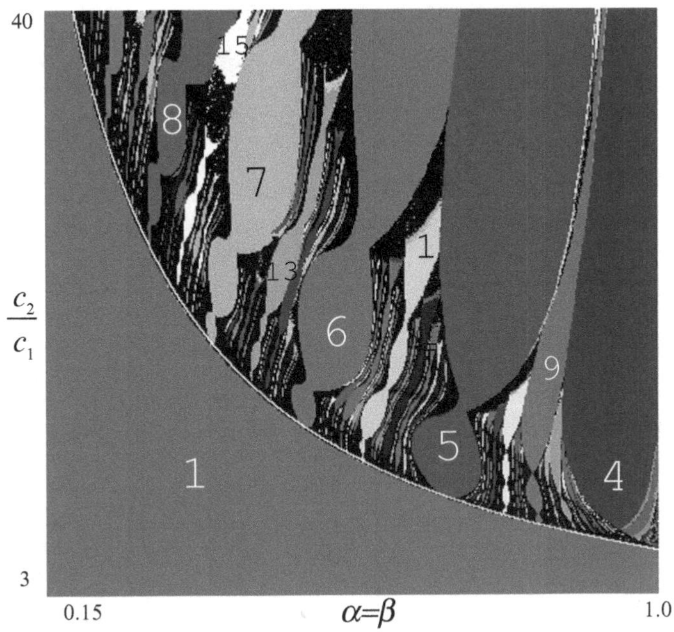

Fig. 3.7 Neimark bifurcation curve and periodic Arnol'd tongues

attractors, for instance, periodic points in a bifurcation cascade. In the subcritical bifurcation, the fixed point continues to coexist with, for instance, a periodic attractor, and eventually loses its stability when its basin contracts around the fixed point.

The supercritical bifurcations are soft, for instance, when a 2-period orbit replaces a fixed point, and the points visited in the oscillatory motion are so close that they can hardly be distinguished from the original fixed point. The subcritical bifurcations are hard; when a system that rested in a fixed point for some while loses its basin of attraction, it has to go to one of the previously coexistent attractors which as a rule consist of points at appreciable distance from the destabilised fixed point.

Shifting focus from phase space to parameter space, subcriticality shows up in the fact that the Arnol'd like periodicity tongues penetrate *through* the bifurcation curve.

The subcritical bifurcations are complicated, because prior to them, the periodic attractors coexisting with the fixed point must have arisen through other bifurcations which are extremely complicated. Agliari et al. (2005a, b) have studied some scenarios for these.

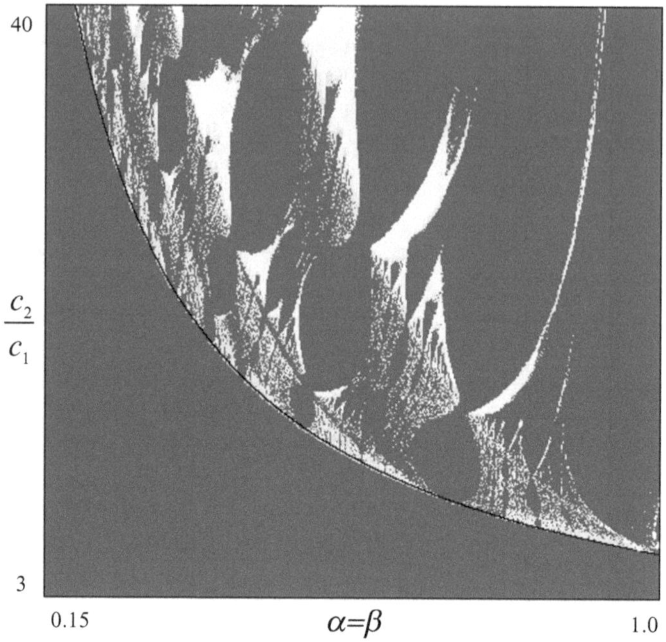

Fig. 3.8 Areas of positive (bright) and negative (dark) Lyapunov exponents

At the moment of subcritical Neimark bifurcation the Arnol'd like periodicity tongues protrude through the bifurcation curve, and extend over entire arcs on this curve.

Though the Arnol'd like tongues do not grow out from the Neimark bifurcation curve, they can still be roughly located through substituting various rational multiples of 2π for θ in the eigenvalues. So, substitute:

$$\theta = \frac{2\pi m}{n} \tag{3.42}$$

in (3.40), and solve for:

$$\alpha + \beta = 2\left(1 - \cos\frac{2\pi m}{n}\right) \tag{3.43}$$

One could in stead use (3.41), but the result would have been the same. Putting $m = 1$, one gets: $\alpha + \beta = 0$ for $n = 1$, $\alpha + \beta = 4$ for $n = 2$, $\alpha + \beta = 3$

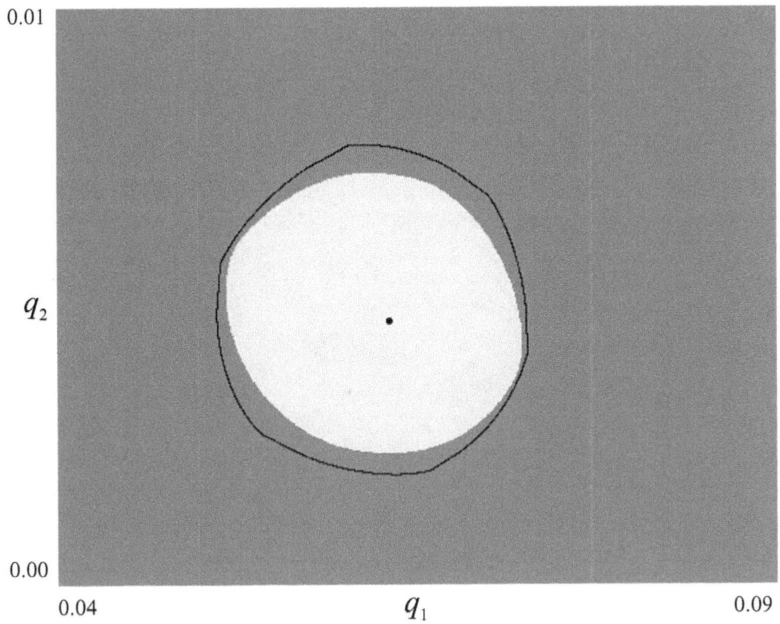

Fig. 3.9 Coexistent fixed point and invariant curve. Parameters $\alpha = \beta = 0.5$, $c_2 / c_1 =$ 13.7635

for n = 3, and $\alpha + \beta = 2$ for n = 4. This indicates that the lowest resonance one can find is with $n = 4$, as otherwise the adaptation coefficients sum up to more than possible when they are in the unit interval.

The following higher harmonics are $\alpha + \beta = \frac{1}{2}\left(5 - \sqrt{5}\right) \approx 1.382$ for n = 5, and $\alpha + \beta = 1$ for $n = 6$, which are perfectly located in the admissible parameter range.

The bifurcation diagram can be seen in Fig. 3.7, and confirms that the corresponding resonance tongues indeed are located as expected from the above calculations.

However, note that there are three parameters, the marginal cost ratio, c_2 / c_1, and the adjustment speeds α and β. In order to make it possible to draw a plane diagram, their number must be reduced to two. Numerical experiment indicates that the best choice is to put $\alpha = \beta$, and to use this along with the marginal cost ratio c_2 / c_1.

In the picture are identified, from right to left, the tongues of periodicities n = 4 and higher, quite as one can infer from the above calculations. The

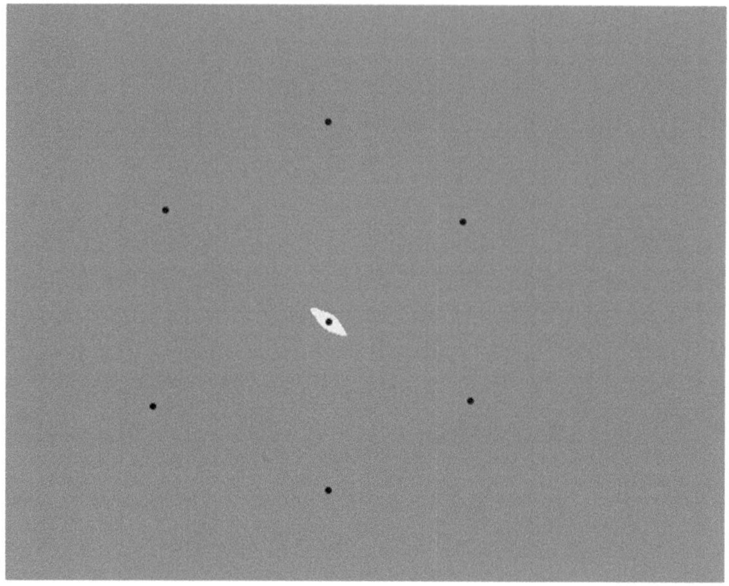

Fig. 3.10 Coexistent fixed point and 6-period orbit. Parameters $\alpha = \beta = 0.5$, $c_2 / c_1 =$ 13.9245

biggest tongues are indicated by numbers. There is one such series running from 4 to 8, the higher ones are too thin for numbering. However, there are also larger tongues numbered, 9, 11, 13, and 15, interspersed between the tongues already spoken of.

The reason for this is that in the above calculations there was a fixed $m = 1$, so the resonances written in full were 1:4, 1:5, 1:6, etc. But these are the same as 2:8, 2:10, 2:12, etc., so it is not surprising that one can find resonances 2:9, 2:11, etc. in between. With equal adjustment speeds, (3.43) reads $\alpha = \beta = 1 - \cos(2\pi m / n)$, so for m = 1, n = 4 one has $\alpha = \beta = 1$, and for $m = 1, n = 5,\ \alpha = \beta = 1 - \cos(2\pi / 5) \approx 0.691$.

Calculating $\alpha = \beta = 1 - \cos(4\pi / 9) \approx 0.826$ for $m = 2$, $n = 9$, it is obvious that indeed the 2:9 tongue fits between the 1:4 and 1:5 tongues.

In all, the 45 lowest periods were checked in Fig. 3.7. As the tongues get thinner and thinner, more than this is difficult to visualize. Note that the periodicity tongues penetrate through the bifurcation curve, as expected with subcriticality.

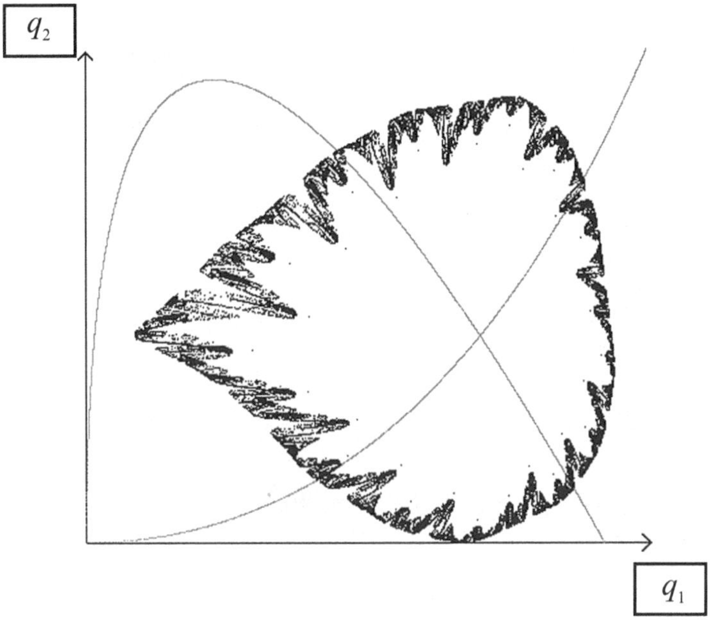

Fig. 3.11 Fractal attractor of the adaptive model *Parameters:* $\alpha = \beta = 0.9$, $c_2 / c_1 = 6.9262$

A companion to Fig. 3.7, is Fig. 3.8, which displays the areas of positive Lyapunov exponents in the parameter plane in bright shade. Note how this picture of the presumably chaotic regions fits into the areas where no periodic tongues show up in Fig. 3.7. Note that the area for positive Lyapunov exponets, quite as the periodicity tongues, protrudes *through* the Neimark bifurcation curve.

In Figs. 3.9 and 3.10 the coexistence of the fixed point with another attractor is shown just before the Neimark bifurcation. As shown, the 6-period tongue starts out at the parameter value $\alpha = \beta = 0.5$.

To get the corresponding value for the marginal cost ratio, substitute this in (3.34), and solve for $c_2 / c_1 = 4\sqrt{3} + 7 \approx 13.9282$.

The stable fixed point for parameters just under the bifurcation curve, coexists with another attractor. So, for Fig. 3.9 choose a slightly lower c_2 / c_1 =13.7635, which is just after a pair of invariant curves, one stable and another unstable, have been born. The stable curve starts getting a hexagonal feature as we are in the 6-period range, the unstable one is the boundary of the basin of attraction for the fixed point (bright shade).

Figure 3.10 is drawn much closer to the Neimark bifurcation curve, where $c_2 / c_1 = 13.8245$. The basin of attraction for the fixed point has now shrunk together around the fixed point, and will the next moment annihilate its stability. Moreover, the closed invariant curve has undergone a global saddle-node bifurcation and broken up to a 6-period cycle, as indicated by Fig. 3.7, where the 6-period tongue penetrates through the bifurcation curve.

Finally, there is on display a chaotic attractor in the shape of a "leaf" in Fig. 3.11, where there is no longer any coexistence with a stable fixed point. The parameter values are $\alpha = \beta = 0.9$, $c_2 / c_1 = 6.9262$.

Analysis by Critical Lines

To analyse the shape of a chaotic attractor such as in Fig. 3.11, the method of critical lines is a most efficient tool. See Abraham et al. (1997) and Mira et al. (1996) The method is based on the fact that the images of points and point sets by a non-invertible map are folded in phase space; on one side of the fold there may, for instance, be no pre-images of points at all, i.e., points mapped there by one iteration, whereas on the other side there may be two different that are mapped into the same point.

These lines or curves where space is folded, mathematically, where the Jacobian vanishes, provide the organizing element for the method. A suitable arc of such a fold line, called critical line, when iterated over and over, produces a train of arcs meeting under tangency, which eventually closes up, encircling an area from which the orbits never can escape. This area is called the absorbing area; unlike the basin of attraction, it is closely wrapped around the attractor, so as to give an accurate picture of its shape, even including interior holes.

There is no need to go into detail of the method here, it is sufficient to see how it works. So, put the determinant of the Jacobian matrix (3.32) equal to zero. Observe that one needs the general matrix (3.32), *not* the one at the Cournot point (3.33). Further note that the determinant is put equal to *zero*, not to *unity* as in deriving the Neimark bifurcation curve (3.34), the purpose being to locate the curve along which the map folds the plane.

So, putting the determinant of (3.32) equal to zero one obtains,

$$\left(\frac{1}{2\sqrt{c_1 q_2(t)}} - 1 \right) \left(\frac{1}{2\sqrt{c_2 q_1(t)}} - 1 \right) = \frac{(1-\alpha)(1-\beta)}{\alpha\beta} \tag{3.44}$$

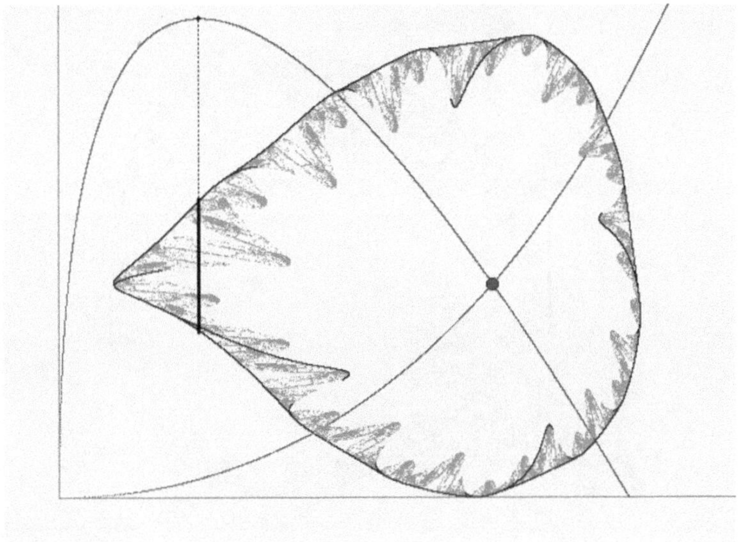

Fig. 3.12 Chaotic attractor and absorbing area with 17 critical lines

It is easy to see that this implicit equation represents a pair of hyperbolas in the q_1, q_2 -plane. Their exact forward images according to (3.28) and (3.29) are too complex to write down, but they can be easily traced by the computer.

Further, the numerical case illustrated in Fig. 3.11 has $\alpha = \beta = 0.9$. Hence the right hand side of (3.44) is small, of the order of magnitude 0.01. If it were zero then either of the brackets on the left would have to be zero. The pair of hyperbolas then becomes a cross of straight lines. This is almost true also with small but nonzero right hand sides, at least with a certain distance from the intersection of this "cross" (the hyperbola asymptotes).

In Figs. 3.12-3.13 the relevant part of L_{-1}, i.e., that which intersects the attractor, almost is a vertical line segment as far as we can see in these pictures. It corresponds to the second bracket in (3.44) being almost zero, i.e., to $q_1 \approx \frac{1}{4c_2}$. As noted before, this coordinate value represents the maximum of the reaction function for the second firm.

In Fig. 3.12 are further shown 17 of the forward iterates of this segment, which obviously suffice to outline the outward boundary of the attractor. This is the case illustrated in Fig. 3.11, and the outline of the attractor can be easily recognized. It has been superposed in gray shade. Further, the reaction functions and the unstable Cournot point at their intersection are displayed.

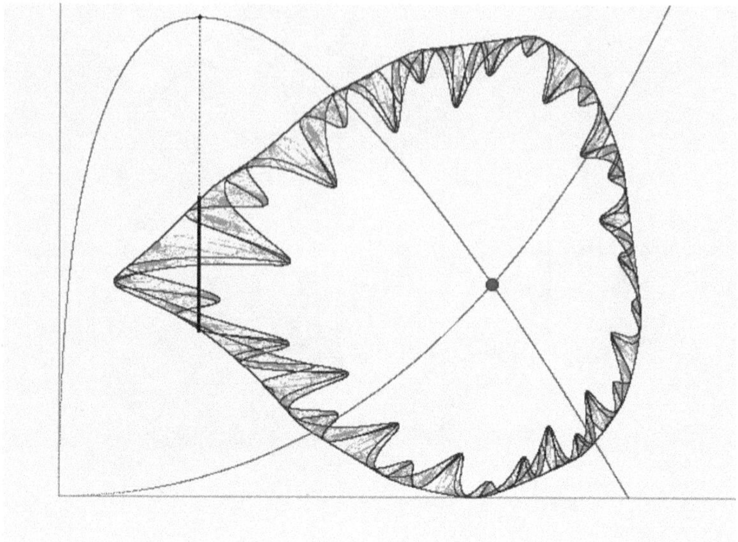

Fig. 3.13 Chaotic attractor and absorbing area with 55 critical lines

By using 55 critical lines as illustrated in Fig. 3.13, the inner boundary of the attractor as well, and other structures, such as densifications of attractor points can be identified.

Even more accurate and efficient would have been to exclude the middle part of the L_{-1} segment displayed, which, as we see, coincides with the "hole" in the attractor, but we skip this computational complication.

References

Abraham, R. H., Gardini, L., & Mira, C. (1997). *Chaos in Discrete Dynamical Systems*, Heidelberg: Springer.

Agliari, A., Gardini, L., & Puu, T. (2005a). Global bifurcations in duopoly when the Cournot point is destabilized through a Subcritical Neimark bifurcation. *International Game Theory Review*, *8*, 1–20.

Agliari, A., Gardini, L., and Puu, T. (2005a, b). Some global bifurcations related to the apperance of closed invariant curves. *Mathematics and Computers in Simulations*, *68*, 201–219.

Bischi, G. I., Mammana, C., & Gardini, L. (2000). Multistability and cyclic attractors in duopoly games. *Chaos, Solitons & Fractals*, *11*, 543–564.

Cournot, A. (1838). *Récherces sur les principes mathématiques de la théorie des richesses.* Paris: Dunot.

Mira, C., Gardini, L., Barugola, A., & Cathala, J. C. (1996). *Chaotic Dynamics in Two-Dimensional Noninvertible Maps,* Singapore: World Scientific.

Puu. (1991, 2004). Chaos in duopoly pricing. *Chaos, Solitons & Fractals, 1*, 573–581, republished in: J.B. Rosser (Ed.), Complexity in Economics: The *International Library of Critical Writings in Economics,* 174. Chelterrhan: Edward Elgar.

Rand, D. (1978). Exotic phenomena in games and duopoly models. *Journal of Mathematical Economics, 5*, 173–184.

Tramontana, F., Gardini, L., & Puu, T. (2010). New properties of the Cournot duopoly with isoelastic demand and constant unit costs, Chaos, Solitons, & Fractals. (in press).

4 Stackelberg Leadership

Stackelberg and Cournot

About a Century after Cournot invented duopoly theory (1838), Heinrich von Stackelberg (1934) proposed one of the most original developments of it.

Cournot's theory was based on the concept of reaction functions. These represent the calculated best replies to each and every possible move by the competitors. The replies are obtained through partial profit maximisation, taking the competitors' supplies as given. When planning for the future, only *expectations* concerning the competitors' supplies, based on *past* experience, can be formed, but, like most expectations, they can turn out wrong as the real process unfolds.

When all competitors form expectations from past experience and take actions for the future, a *dynamic* process is formed. In the simplest case, each competitor just expects the others to stay with their previous moves. In this way the reaction functions get a dynamic interpretation, and become the basis for a recursive dynamic system or map. The reaction functions may also be used as simultaneous to define an *equilibrium* when they all hold at once; the Cournot equilibrium. Both interpretations can be said to be implicit in Cournot's original work.

Stackelberg's work, restricted to duopoly (two competitors), extended Cournot's model, as he proposed that either competitor could *learn* the true reaction function of the other, hence becoming a "leader" treating the other as "follower", and so solve, not a partial optimization problem conditional upon the expected decision by the competitor, but an optimization problem that takes explicit account of all such partial optimization reactions by the competitor.

If the other competitor actually behaves as assumed, i.e., follows the proper Cournot reaction function, then a consistent leader/follower pair can be formed. This possibility exists for both competitors.

T. Puu, *Oligopoly: Old Ends – New Means*, DOI 10.1007/978-3-642-15964-0_4,

In this way two Stackelberg equilibria, with each duopolist as potential leader, are added to the Cournot equilibrium. If both competitors behave as followers the Cournot case is retrieved. It is hence clear that Stackelberg's theory belongs to the Cournot setting. In this sense Stackelberg's solution generalises Cournot's. But it is more restricted in the sense that it only comes in a static version.

If one wants to fit the models together in one dynamic model, it is clear that it is the Cournot format that is suitable for extension, so that it can accommodate the Stackelberg equilibria along with the Cournot equilibrium as fixed points. But, it is not obvious how such dynamisation should be done.

The problem is that in the *long run* Stackelberg's *equilibrium* is more profitable for the leader than is Cournot's equilibrium. But, in the *short run* it is always profitable for a Stackelberg leader to temporarily jump to Cournot action, even when being successful in the sense that the competitor actually behaves as a follower. The simple reason for this is that Cournot's reaction function is *defined* as the best reply in each situation. Such switching, however, is no deal for more than a short time. Already in the period after it would have been better to stay in Stackelberg equilibrium.

These things make the formulation of a combined dynamic more tricky than might first be expected. One may note that, a bit surprisingly, Stackelberg never took any dynamic perspective, though it seems that if the follower always keeps to the reaction function, the leader might also devise a changing supply policy over time in the pursuit of maximum leadership profit. It can be shown (see Puu 2010) that an oscillating leadership sales policy would in fact result in higher profit than sticking to an equilibrium solution constant over time.

How can Stackelberg action be included in the dynamic Cournot setting? For the dynamic system, one needs an endogenous condition for jumping between Cournot to Stackelberg action. Such a clause is most conveniently based on expected profits from Cournot best reply compared to Stackelberg leadership profits in each temporary situation.

But, the clause for choosing Stackelberg leadership cannot be based on Stackelberg profits exceeding temporary Cournot action profits, because even in successful leadership, reversion to temporary Cournot action *always* results in a higher profit. A Stackelberg equilibrium could therefore never become a stable equilibrium for the Cournot map.

The comparison therefore needs to incorporate a weighting factor prescribing, for instance, a switch to Stackelberg action whenever Stackelberg profits exceed, say 75 or 50, not 100% of current expected Cournot profits.

In the long run such a rule might result in a good average profit. This model setup was recently proposed in Puu (2009).

Recapitulation of the Cournot Model

It is now high time to make these things precise in terms of a specific model. The case with isoelastic demand and constant marginal costs, discussed in Chap. 3, is convenient also for the present purpose. A short recapitulation may be convenient. The respective profits were

$$\Pi_1(t+1) = \frac{q_1(t+1)}{q_1(t+1)+q_2(t)} - c_1 q_1(t+1) \tag{4.1}$$

$$\Pi_2(t+1) = \frac{q_2(t+1)}{q_1(t)+q_2(t+1)} - c_2 q_2(t+1) \tag{4.2}$$

Note that the variables are now dated. The difference in time period attribution is due to the fact that each firm's own supply is the one planned according to the reaction function, whereas the competitor's is the expected (equal to that of the preceding period). The Cournot reaction functions maximizing these profits were

$$q_1(t+1) = \begin{cases} \sqrt{\dfrac{q_2(t)}{c_1}} - q_2(t) & q_2(t) < \dfrac{1}{c_1} \\ 0 & q_2(t) \ge \dfrac{1}{c_1} \end{cases} \tag{4.3}$$

$$q_2(t+1) = \begin{cases} \sqrt{\dfrac{q_1(t)}{c_2}} - q_1(t) & q_1(t) < \dfrac{1}{c_2} \\ 0 & q_1(t) \ge \dfrac{1}{c_2} \end{cases} \tag{4.4}$$

It is also convenient to recall the facts about Cournot equilibrium from Chap. 2. Its phase coordinates were

$$q_1 = \frac{c_2}{\left(c_1 + c_2\right)^2} \tag{4.5}$$

$$q_2 = \frac{c_1}{\left(c_1 + c_2\right)^2} \tag{4.6}$$

and the resulting Cournot equilibrium profits

$$\Pi_1 = \frac{c_2^{\;2}}{\left(c_1 + c_2\right)^2} \tag{4.7}$$

$$\Pi_2 = \frac{c_1^{\;2}}{\left(c_1 + c_2\right)^2} \tag{4.8}$$

Note that (4.7) and (4.8) are the equilibrium profits. For setting up the mixed process, also the temporary Cournot reaction profits will be needed. To obtain these, substitute from the main branches of the reaction functions (4.3) and (4.4) into the profit expressions (4.1) and (4.2). Thus

$$\Pi_1(t+1) = \left(1 - \sqrt{c_1 q_2(t)}\right)^2 \tag{4.9}$$

$$\Pi_2(t+1) = \left(1 - \sqrt{c_2 q_1(t)}\right)^2 \tag{4.10}$$

It might be interesting to check under which circumstances these temporary profits are higher than those in Cournot equilibrium. Hence, compare (4.9) to (4.7) and (4.10) to (4.8). As all right hand sides in (4.7)–(4.10) are squared, the conditions simplify to

$$1-\sqrt{c_1 q_2(t)} > \frac{c_2}{c_1+c_2}$$

$$1-\sqrt{c_2 q_1(t)} > \frac{c_1}{c_1+c_2}$$

respectively. Rearranging, and squaring to get rid of the root signs, the expressions simplify to

$$q_2(t) < \frac{c_1}{(c_1+c_2)^2} \tag{4.11}$$

$$q_1(t) < \frac{c_2}{(c_1+c_2)^2} \tag{4.12}$$

The right hand sides of (4.11) and (4.12) are identical to the respective Cournot equilibrium coordinates as stated in (4.5) and (4.6), so the conditions for temporary profits exceeding Cournot equilibrium profits have very simple interpretations. In particular, any time the process unfolds in the rectangle of phase space whose upper right corner is the Cournot point, both duopolists are better off than in that equilibrium point.

Stackelberg Equilibria

Firm 1 Leader. In the case of Stackelberg action the would-be leader, say the first firm, learns to know the second firm's reaction function (4.4), and takes advantage of it, substituting all reactions in its proper profit function (4.1). One takes the main branch, because zero supply from the competitor would make the leader monopolist, and this was to be avoided. As stated above, Stackelberg's theory only came in the equilibrium format, so there is no need to date any phase variables. The substitution hence results in

$$\Pi_1 = \frac{q_1}{q_1 + \sqrt{\frac{q_1}{c_2}} - q_1} - c_1 q_1 = \sqrt{c_2 q_1} - c_1 q_1 \tag{4.13}$$

which is a function q_1 alone. It is now possible to put the *total* derivative equal to zero, and solve, to obtain

$$q_1 = \frac{c_2}{4c_1^2} \tag{4.14}$$

as leadership supply. The corresponding follower supply is obtained from the reaction function (4.4), main branch, again skipping period attributions

$$q_2 = \sqrt{\frac{1}{c_2}\frac{c_2}{4c_1^2}} - \frac{c_2}{4c_1^2} = \frac{2c_1 - c_2}{4c_1^2} \tag{4.15}$$

It is obvious that this Stackelberg solution only makes sense if $c_2 < 2c_1$. Using (4.15) in (4.1), the leadership profit becomes

$$\Pi_1 = \frac{\frac{c_2}{4c_1^2}}{\frac{c_2}{4c_1^2} + \frac{2c_1 - c_2}{4c_1^2}} - c_1 \frac{c_2}{4c_1^2} = \frac{c_2}{4c_1} \tag{4.16}$$

It is easily verified that leadership profits always exceed Cournot equilibrium profits, i.e.,

$$\frac{c_2}{4c_1} > \frac{c_2^2}{(c_1 + c_2)^2} \tag{4.17}$$

unless $c_1 = c_2$, in which case equality holds; so *in equilibrium* Stackelberg leadership is always better than Cournot action. To find this out, cancel one power of c_2, which is positive, in (4.17), multiply through crosswise by the denominators, also positive, and note that $(c_1+c_2)^2-4c_1c_2=(c_1-c_2)^2>0$.

It should be recorded that this advantage of being a leader holds for both duopolists, no matter if they have the higher or lower production costs.

Firm 2 Leader. For completeness, record these facts for the case when the second firm attempts leadership. Its leadership profits are

$$\Pi_2=\frac{q_2}{\sqrt{\frac{q_2}{c_1}}-q_2+q_2}-c_2q_2=\sqrt{c_1q_2}-c_2q_2 \tag{4.18}$$

which is a function q_2 alone. Thus, optimising,

$$q_2=\frac{c_1}{4c_2^{\,2}} \tag{4.19}$$

is the leadership supply. The follower's supply then becomes

$$q_1=\sqrt{\frac{1}{c_1}\frac{c_1}{4c_2^{\,2}}}-\frac{c_1}{4c_2^{\,2}}=\frac{2c_2-c_1}{4c_2^{\,2}} \tag{4.20}$$

and is relevant as long as $c_1<2c_2$. The leadership profits are

$$\Pi_2=\frac{\frac{c_1}{4c_2^{\,2}}}{\frac{c_1}{4c_2^{\,2}}+\frac{2c_2-c_1}{4c_2^{\,2}}}-c_2\frac{c_1}{4c_2^{\,2}}=\frac{c_1}{4c_2} \tag{4.21}$$

Again

$$\frac{c_1}{4c_2}>\frac{c_1^{\,2}}{(c_1+c_2)^2} \tag{4.22}$$

Stackelberg's original point was that both (4.14)–(4.15) and (4.19)–(4.20) work as consistent equilibrium points. He also noted that if both acted as followers, then the Cournot equilibrium (4.5)–(4.6) would result, and that it is not consistent for both competitors to be leaders at the same time.

Within the present model, it is profitable for both duopolists to attempt leadership, as seen from (4.17) and (4.22), but this only works when $c_2 < 2c_1$ and $c_1 < 2c_2$ hold, i.e., when no firm's unit cost exceeds twice the other's.

How to Fit Stackelberg Action in the Cournot Dynamics

It is now time to check out the possibilities to include Stackelberg action in the Cournot dynamic process. As mentioned, there is something of a paradox. Stackelberg *equilibrium* is always better than Cournot equilibrium, but, as shown in (4.11) and (4.12), profits in the Cournot process can sometimes exceed the equilibrium profits. The question is, when do they exceed *even* Stackelberg leadership profits?

Suppose one firm is *successful* in establishing leadership, the other acting as a follower, and take this as the starting position. For definiteness assume the first firm is the leader. Then it receives the profits stated in (4.16). Its supply is stated in (4.14), and the follower's supply in (4.15).

So what is the best next action for this successful leader. Observe that now there is no uncertainty concerning expectations in this consistent Stackelberg equilibrium.

Here comes the paradox: Even from this advantageous position is it *always* better for the *leader* to switch to Cournot action! The temporary Cournot profit was recorded in (4.9), so substitute the follower's action from (4.15)

$$\Pi_1(t+1) = \left(1 - \sqrt{c_1 \frac{2c_1 - c_2}{4c_1^2}}\right)^2 = \left(1 - \frac{1}{2}\sqrt{2 - \frac{c_2}{c_1}}\right)^2 \tag{4.23}$$

Comparing this to (4.16) one finds that always

$$\left(1 - \frac{1}{2}\sqrt{2 - \frac{c_2}{c_1}}\right)^2 > \frac{c_2}{4c_1} \tag{4.24}$$

unless $c_1 = c_2$. As a conclusion, *it is always profitable for the successful Stackelberg leader to switch to Cournot action*!

However, this advantage only lasts for a short while. In (3.27) it was stated that Cournot equilibrium is stable when $\frac{c_2}{c_1} \in \left]3 - 2\sqrt{2}, 3 + 2\sqrt{2}\right[$, whereas from (4.15) and (4.20) the Stackelberg equilibria exist when $\frac{c_2}{c_1} \in \left]\frac{1}{2}, 2\right[$. Hence, the Cournot equilibrium is stable when the Stackelberg equilibria exist. This means that once a Stackelberg equilibrium is left, a stable Cournot process sets in and converges to the Cournot equilibrium point. But, in the long run this is worse than staying in stationary Stackelberg leadership.

Thus, on one hand, it is always profitable to jump from successful Stackelberg leadership to a dynamic Cournot process, but in the long run this leads to a worse situation than staying in leadership. It goes without saying that this holds for both duopolists.

Obviously, some policy of jumping between different actions is more profitable than staying in either Cournot or Stackelberg equilibria. Hence there is a point in setting up an endogenous branch condition for jumping. Somehow it must be based on comparing expected profits. But, it cannot be based on a comparison of plain expected profits, because the Cournot reaction function was defined as the best response in each situation, so therefore Stackelberg action would never be chosen. It will if one stipulates that the competitors attempt Stackelberg leadership when expected leadership profits exceed 50 or 75% of expected short run Cournot profits; not 100%, because then the condition becomes inoperative. In general assume a balance factor k, such that firm 1 chooses Stackelberg action when

$$\left(1 - \sqrt{c_1 q_2(t)}\right)^2 < k \frac{c_2}{4c_1} \tag{4.25}$$

Likewise, firm 2 chooses Stackelberg action when

$$\left(1 - \sqrt{c_2 q_1(t)}\right)^2 < k \frac{c_1}{4c_2} \tag{4.26}$$

For branch choice conditions in a map, it is more convenient to state these explicitly in terms of the phase variables, i.e.,

$$q_2(t) > \frac{1}{c_1}\left(1 - \frac{1}{2}\sqrt{k\frac{c_2}{c_1}}\right)^2 \tag{4.27}$$

$$q_1(t) > \frac{1}{c_2}\left(1 - \frac{1}{2}\sqrt{k\frac{c_1}{c_2}}\right)^2 \tag{4.28}$$

The Combined Map

Equations (4.27) and (4.28) state the conditions for choosing between Cournot and Stackelberg action, but it is also necessary to recall the conditions for nonnegativity of the Cournot reaction functions as stated in (4.3) and (4.4). To set the branches properly it is interesting to know whether (4.27) is less or more restrictive than (4.3), and the same about (4.28) as compared to (4.4).

Obviously this depends on the squared parentheses in (4.27) and (4.28). They cannot exceed unity, but they may fall short of minus unity. This happens when $k\frac{c_2}{c_1} > 16$ or $k\frac{c_1}{c_2} > 16$ respectively. Recalling that the marginal cost ratios must not exceed 2 if the Stackelberg equilibria are to exist one finds the necessary condition $k > 8$, for the nonnegativity conditions to be the stricter. As this means $1/k < 0.125$, or that the duopolists jump to Stackelberg action whenever leadership profits exceed 12.5% or lower of expected Cournot action profits, the case would be rather extreme.

The resulting three-branch maps are now

$$q_1(t+1) = \begin{cases} \sqrt{\dfrac{q_2(t)}{c_1}} - q_2(t) & q_2(t) < \dfrac{1}{c_1}\left(1 - \dfrac{1}{2}\sqrt{k\dfrac{c_2}{c_1}}\right)^2 \\ \dfrac{c_2}{4c_1^{\,2}} & \dfrac{1}{c_1}\left(1 - \dfrac{1}{2}\sqrt{k\dfrac{c_2}{c_1}}\right)^2 \le q_2(t) < \dfrac{1}{c_1} \\ 0 & \dfrac{1}{c_1} \le q_2(t) \end{cases} \tag{4.29}$$

$$q_2(t+1)=\begin{cases}\sqrt{\dfrac{q_1(t)}{c_2}}-q_1(t) & q_1(t)<\dfrac{1}{c_2}\left(1-\dfrac{1}{2}\sqrt{k\dfrac{c_1}{c_2}}\right)^2\\ \dfrac{c_1}{4c_2^{\,2}} & \dfrac{1}{c_2}\left(1-\dfrac{1}{2}\sqrt{k\dfrac{c_1}{c_2}}\right)^2\le q_1(t)<\dfrac{1}{c_2}\\ 0 & \dfrac{1}{c_2}\le q_1(t)\end{cases} \tag{4.30}$$

Numerical Study

The map (4.29) and (4.30) is too complex for analysing analytically more than was already done with putting up its bits and pieces. However, it is very simple to implement for numerical study with any standard computer software. It should be mentioned that the zeros in the third branches, quite as in the cases without Stackelberg action, present some difficulty. Occasionally, in order to stop the computer from sticking to the origin, a negligibly small "stand by" supply is therefore substituted for the exact zero in these third branches.

The Bifurcation Plane

For simplicity the balancing factor k is taken the same for both duopolists, though it would be more general to differentiate between the competitors. In this way it is possible to display bifurcations in a parameter plane as there also are other crucial parameters c_2, c_1. In the illustrations $c_1 = 0.5$ is kept fixed and the marginal cost ratio c_2 / c_1 is what is displayed.

The bifurcation diagram is shown in Fig. 4.1. For each parameter combination represented by a point in the parameter plane, an initial condition for q_1, q_2 close to the Cournot equilibrium point was taken. After a large number of iterations the asymptotic orbit was identified. Depending on the ultimate periodicity, the point in parameter plane was shaded, and the areas were labelled with numbers indicating the periodicity, 1 of course indicating a fixed point. It should be noted that always starting from a point close to the Cournot point does not disclose whether there is any coexistence of the Cournot equilibrium with some other periodic orbit whose attraction basin is farther away from the Cournot point, but these issues are discussed later on.

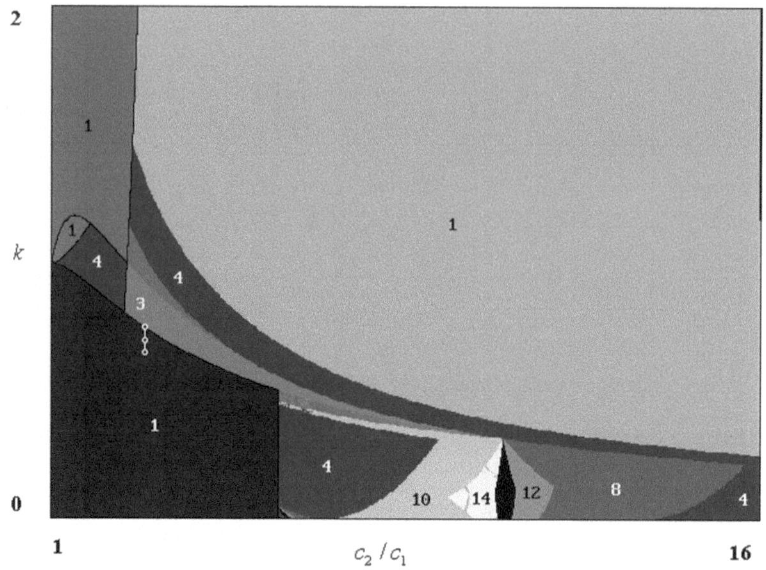

Fig. 4.1 Bifurcation diagram in parameter space. Observable are regions of periodic orbits. There are fixed points (period 1) of four different types, further orbits of periods 3, 4, 8, 10, 12 and 14

Obviously most of the plane contains areas representing period 1 orbits, i.e., attracting fixed points. These are of four different types.

Fixed Points

1. At the bottom of the picture (labelled 1), the Cournot attracting equilibrium area is shown. It yields to periodic orbits where the Cournot equilibrium turns unstable at $c_2 / c_1 = 3 + 2\sqrt{2}$; first to period 4, then period 10, then 14, etc.

2. There is also the tiny lens shaped attracting Stackelberg area (also labelled 1) for one leader/follower pair half way up to the left. Note that this picture was drawn starting at $c_2 / c_1 = 1$. In the interval $[0,1]$ there is a distorted and reflected image of the facts displayed for the interval $[1,\infty]$.

3. Further, there is the region on top left of the picture (labelled 1), with the cuspoid bottom line, representing a nonstandard Stackelberg case where both competitors insist on sticking to Stackelberg leadership (which Stackelberg in the equilibrium format considered unresolvable). True, both competitors get disappointed, as the assumption that the other always acts as a follower does not hold, but under the defined map and the parameter combinations displayed, neither could find expected current profits from Cournot action more profitable.

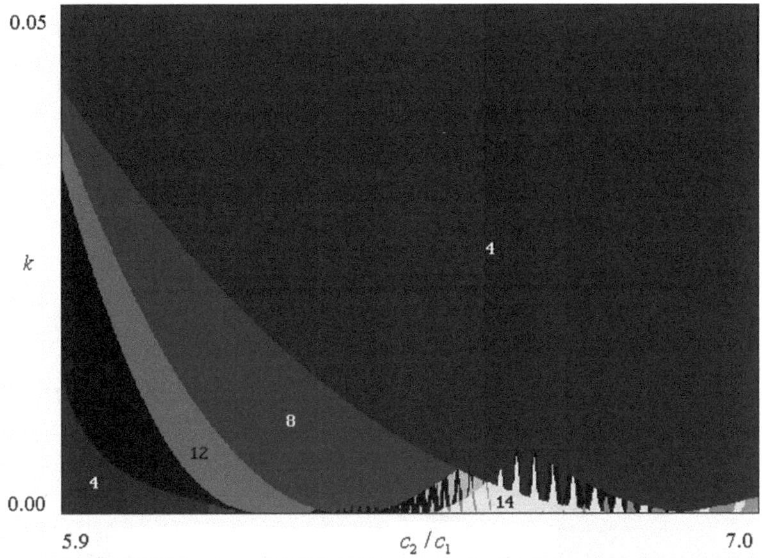

Fig. 4.2 Blowup picture of part of Fig. 4.1 to show more of periodic orbits and overlaps. The *black regions* indicate higher periodicities or more complex dynamics

4. Finally, the region in the upper right part (marked with a period 1 label) represents the case where one competitor sticks to Stackelberg leadership, whereas the other finds no better action than to suppress production altogether, a kind of monopoly. As mentioned, the model with an isoelastic demand function is not suitable to treat monopoly or collusion. This case will therefore not be considered in the sequel.

Periodic Orbits

As so much of the area in Fig. 4.1 is taken up by fixed point regions, little space is left for other attracting orbits. One can see periodic orbits, labelled 3, 4, 8, 10, 12, and 14. This is at the chosen part of parameter plane and the resolution of the picture. No higher periods were checked by the computer. The black area may contain higher periodicities, quasiperiodicity, or even chaos.

There seems to be some irregularity in the lower part of Fig. 4.1 in the interval $c_2 / c_1 \in [6,7]$, so we take a close up picture of it in Fig. 4.2. Notable are the two humps of a spiky appearance. The spikiness indicates coexistence of attractors; for some parameter combinations the chosen initial condition leads to one periodicity, for a nearby combination to another. As we will be shown, the model is proficient in such coexistent attractors

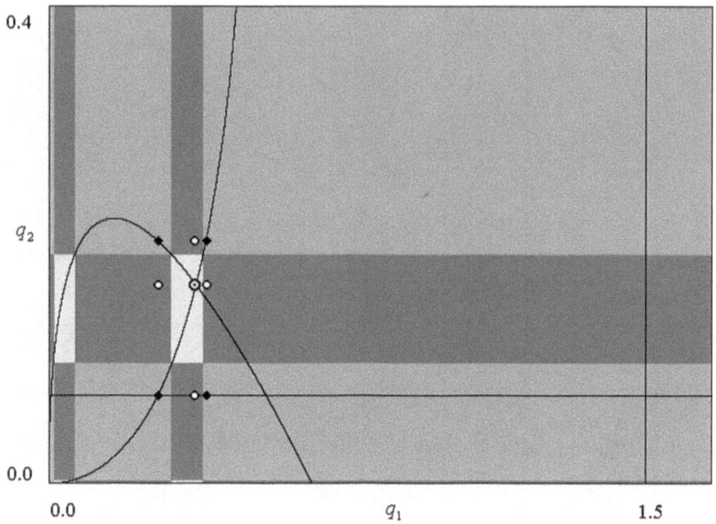

Fig. 4.3 Cournot equilibrium coexisting with two 4-period cycles when $c_1 = 0.5,\quad c_2 = 1.5,\quad k = 0.65$

The Phase Plane

There are many potentially interesting bifurcation scenarios to pursue. Just pick one, with $c_1 = 0.5, c_2 = 1.5$ fixed, and a slightly varying $k \in [0.65, 0.75]$, indicated by the tiny line segment in Fig. 4.1. In Figs. 4.3–4.6, the phase diagram is displayed for $k = 0.65$, $k = 0.7$, $k = 0.745$, and $k = 0.75$.

The phase diagrams show the two Cournot reaction curves, along with the lines representing Stackelberg leadership action, further the different attractors, and their basins of attraction. Note that the periodic orbits are located on the intersection points of a grid of three or four horizontal and vertical lines, hence producing 9 or 16 attractor points. These are distributed between orbits of different periodicities, such as $1+3+6+6=16$, or $3+6=9$, or $1+4+4=9$. The pictures will be discussed one by one, and also the average profits over each cycle will be recorded.

The Cournot equilibrium does not depend on the parameter k, so its coordinates are fixed, and the profits can be calculated from (4.7) and (4.8).

$$\Pi_1 = 0.562$$
$$\Pi_2 = 0.062$$

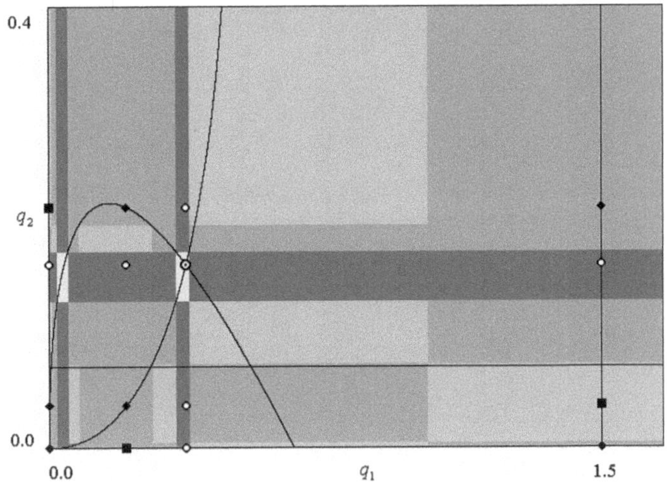

Fig. 4.4 Cournot equilibrium coexisting with one 3- and two 6-period cycles when $c_1 = 0.5, \quad c_2 = 1.5, \quad k = 0.70$.

Notably, Cournot equilibrium profits for the firm with higher marginal costs are only about one tenth of what the firm with lower costs receives. Further, note that outright Stackelberg equilibrium for the low cost firm is not feasible because the cost ratio is too uneven ($c_2 / c_1 = 3 > 2$). This does not prevent the competitors from occasionally attempting Stackelberg action.

As for the periodic orbits, start with taking a look at Fig. 4.4, where $c_1 = 0.5, c_2 = 1.5, k = 0.70$. As the fixed Cournot point is destabilised only at $k = 0.75$, it remains stable. Its basin is the tiny rectangle around the intersection of the Cournot reaction curves, and a few other tiny rectangles in the same shade. The cross in the middle of which the main basin for the Cournot point lies, also contains the points of a 6-period orbit.

In addition there is another 6-period orbit whose points lie on the Cournot reaction function (including the Stackelberg lines). Finally, there is also a 3-period orbit. So, 1 + 3 + 6 + 6 = 16, quite as suggested. The average profits per period for these cases are

$$\Pi_1 = 0.377$$
$$\Pi_2 = 0.240$$

$$\Pi_1 = 0.446$$
$$\Pi_2 = 0.172$$

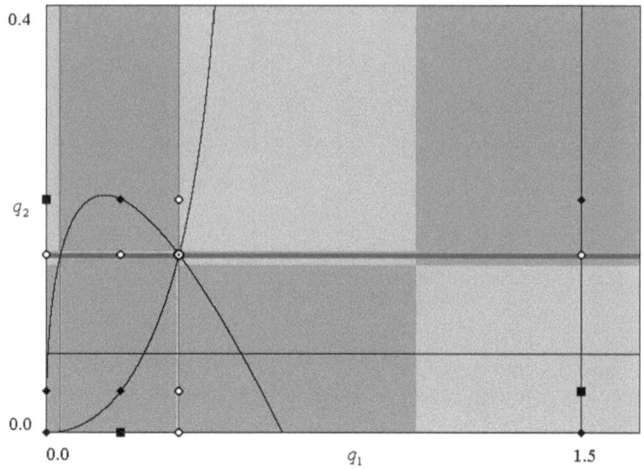

Fig. 4.5 Coexistence of Cournot equilibrium, 3-period cycle, and two 6-period cycles with $c_1 = 0.5, c_2 = 1.5, k = 0.745$

$$\Pi_1 = 0.461$$
$$\Pi_2 = 0.160$$

respectively. It is interesting to note that it is particularly profitable for the firm facing the *higher* marginal cost to try to break out from Cournot equilibrium to a cyclic solution, as this alters the profit shares to its advantage; especially in the 3-period cycle, where the profits for the duopolists have the same order of magnitude.

Next, consider Fig. 4.5, where $k = 0.745$, and the cross is very tiny. The basin of the Cournot point has shrunk to almost nothing, as has the cross itself, which provides the basin for one of the 6-period orbits. The same orbits remain as in the previous picture, though two of them are about to disappear.

This indeed happens at $k = 0.75$, the subcritical bifurcation in Fig. 4.6, where both the Cournot equilibrium, and one of the 6-period cycles have disappeared. What then remains are two attractors; the 3-period orbit, and a 6-period cycle. In all, there are now 3 + 6 = 9 grid points in the intersections of three (horizontal and vertical) lines. For further increasing k, the 3- and 6-cycles remain.

As for profits, the situation remains the same as before, their entries are not even changed with the chosen number of significant digits, though in the latter case the Cournot equilibrium and the 6-cycle within the cross no longer exist.

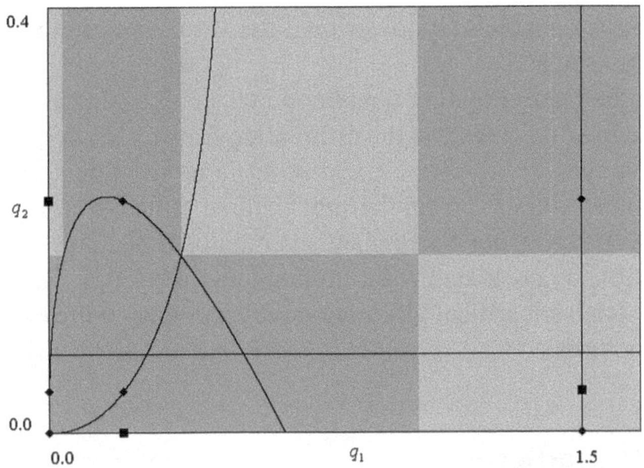

Fig. 4.6 The subcritical bifurcation, only the 3-period cycle, and a 6-period cycle remain at $c_1 = 0.5, c_2 = 1.5, k = 0.75$

Finally, Fig. 4.3 displays the case $k = 0.65$. The higher Stackelberg line is not yet visited, so there remain two cycles, now of period 4, one within the cross, and another outside, with points on the reaction functions. Further, the Cournot point is, of course, also stable. In all, there are $1 + 4 + 4 = 9$ periodic points on a grid of three by three lines.

Now the profit situation is different; the Cournot profits remain the same, but for the 4-period orbit on the reaction functions

$$\Pi_1 = 0.601$$
$$\Pi_2 = 0.074$$

and for the 4-period orbit off these

$$\Pi_1 = 0.579$$
$$\Pi_2 = 0.070$$

are obtained. Now it is profitable for the firm facing lover costs to break up from Cournot equilibrium.

For k increasing from below, this situation emerges at about $k \in [0.61, 0.62]$. For lower k, there is just the Cournot equilibrium point and one basin. It seems that the two 4-period orbits arise simultaneously at a bifurcation for some critical k, which is considerably lower than the value at which the Cournot equilibrium is destabilised.

The fact that the Cournot point, contained in a small basin, remains along with other attractors indicates that the bifurcation from Cournot point to periodic orbits is subcritical, i.e., the fixed point is not just destabilised and replaced by another attractor as in the supercritical bifurcation. It rather disappears through its basin contracting and eventually vanishing, so that other attractors that already coexisted with it remain the only ones. The scenarios connected with such subcritical bifurcations often display quite complicated bifurcation structures.

Learning Periodicities

The question now arises if the duopolists could learn the periodicity they produce. The scenarios checked with coexistent attractors are unnecessarily complicated, so check the facts at a parameter point $c_1 = 0.5, c_2 = 3, k = 0.4$ within the bigger 4-period tongue in Fig. 4.1, where there is just one 4-period cyclic attractor, and a unique basin.

The four phase plane points of the orbit, shown in Fig. 4.7, are as follows:

- **A** The second competitor has previously chosen Stackelberg leadership, the first responds with Cournot action.
- **B** The second competitor reverts to Cournot action.
- **C** The first competitor again responds with Cournot action.
- **D** The second competitor now chooses Stackelberg action, and then everything starts all over again.

Note that point **D** is *not* located on the first firm's reaction function; it is just to the right of it on the second firm's Stackelberg line below point **C**.

Given the simple regularity, suppose the agents learn the periodicity, and react, *not* to the competitor's action *one* period back, but *four* periods back. What would be the outcome?

If one only considers the dynamic every fourth period, the outcome is exactly as before. But, in the three intervening periods, the system will settle to an independent cycle of the same type.

The total result is a composition of four 4-cycles successively displaced in time, and, even if there is just one original cycle, it is *impossible* to make the composition break down to a 4-cycle. The outcome is a 16-period cycle. To see the point, consider the single 4-cycle, whose points are ABCD.

Try to again arrange a four cycle. In the first line of the table there are three blanks between the entries. To produce a sequence beginning ABCD,

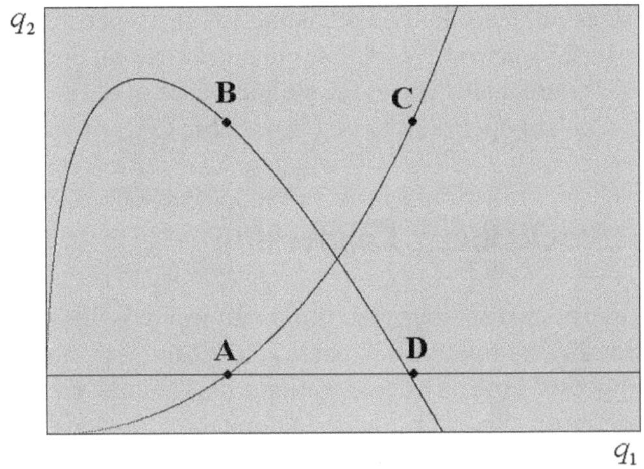

Fig. 4.7 4-period cycle when $c_1 = 0.5, c_2 = 3, k = 0.4$

choose BCDA for the first blanks, CDAB for the second, and DABC for the third, thus completing the sequence.

A				*B*				*C*				*D*			
A	*B*			*B*	*C*			*C*	*D*			*D*	*A*		
A	*B*	*C*		*B*	*C*	*D*		*C*	*D*	*A*		*D*	*A*	*B*	
A	*B*	*C*	*D*	*B*	*C*	*D*	*A*	*C*	*D*	*A*	*B*	*D*	*A*	*B*	*C*

So what is the outcome? The resulting sequence in the bottom row indeed starts with ABCD, but it is not repeated, because then follows BCDA, CDAB, and DABC. Only the full 16-sequence of the bottom line is repeated.

So, if the agents learn the periodicity of 4, they actually produce a periodicity of 16. The reader can try to arrange the sequences differently, but the outcome is always the same; a 16-period cycle.

As the first four entries in the sequence of 16 determine all the following, according to the order established in the original 4-cycle, and as there are four choices for each entry, there would now seem to exist $4^4 = 256$ different cycles. However, it is elementary, but tedious, to check that many are identical, and only 16 are different.

Note that average profits in the 16-period cycle remain the same as in the 4-period cycle, because the same points ABCD are visited with the same relative frequency as before though under different permutations.

To sum up; starting with one 4-period cycle and one basin, learning produces 16 different 16-period cycles, and, of course, 16 different basins.

And so it goes on, learning and adjusting to the 16 periodicity produces a host of different 256-period cycles. The conclusion is that the idea of "rational expectations" is untenable even in the simplest case of only one original orbit of low order. The only periodicity that is possible to learn is a fixed point.

Dynamic Stackelberg Leadership

Stackelberg considered stationary solutions alone, but is this really necessary? If the follower always follows the reaction function, why not attempt a sales policy varying over time? The leader might obtain an even higher profit by designing a periodic sales profile and dragging the follower along all the time, than through keeping to a stationary leadership policy. It seems that under most nonlinear demand functions variable sales policy indeed yields higher average profits than a traditional constant Stackelberg leadership action.

Stackelberg's Demand

Stackelberg used two different exemplifying demand functions; a linear and an exponential. As usual with the restricted class of linear functions, more interesting results fail to show up. However, the exponential in inverse form

$$p = e^{-Q} \tag{4.31}$$

where p denotes price and Q denotes quantity, is well suited to illustrate the point made above. To be quite true, Stackelberg used form

$$p = e^{-\sqrt{Q}} \tag{4.32}$$

including a square root sign. This seems both unnecessarily messy to handle, and a bit contrived, but the reason for its inclusion is easy to see. Like several other duopoly theorists, Stackelberg assumed variable production costs to be zero. Then the competitors would maximize revenues. Assume two competitors supplying q_1 and q_2 respectively. Then, as $Q = q_1 + q_2$, according to (4.31)

$$p = e^{-q_1 - q_2} \tag{4.33}$$

The competitors would attempt to maximize their profits (i.e., revenues),

$$\Pi_1 = e^{-q_1-q_2} q_1 \tag{4.34}$$

$$\Pi_2 = e^{-q_1-q_2} q_2 \tag{4.35}$$

The first order conditions then read

$$(1-q_1)e^{-q_1-q_2} = 0 \tag{4.36}$$

$$(1-q_2)e^{-q_1-q_2} = 0 \tag{4.37}$$

returning the reaction functions

$$q_1 = 1 \tag{4.38}$$

$$q_2 = 1 \tag{4.39}$$

As a consequence these constant reaction functions make the Cournot equilibrium and the Stackelberg equilibria identical, so, using (4.31), Stackelberg would have missed his point. The square root in (4.32) saved him from that.

Cournot Reaction Functions

However, the same result is obtained through including variable costs, which for simplicity can be taken as constant, c_1 and c_2 per unit output for the two competitors, and then the square root sign, for which there is no particular reason, and which complicates the derivations, can just be omitted.

Profits are then

$$\Pi_1 = \left(e^{-q_1-q_2} - c_1\right) q_1 \tag{4.40}$$

$$\Pi_2 = \left(e^{-q_1-q_2} - c_2\right) q_2 \tag{4.41}$$

and first order maximum conditions now read

$$(1-q_1)e^{-q_1-q_2} = c_1 \tag{4.42}$$

$$(1-q_2)e^{-q_1-q_2} = c_2 \tag{4.43}$$

Equations (4.42) and (4.43) cannot be solved in closed form for the Cournot reaction functions, but they can for the inverses

$$q_2 = \phi_1^{-1}(q_1) = \ln\frac{1-q_1}{c_1} - q_1 \tag{4.44}$$

$$q_1 = \phi_2^{-1}(q_2) = \ln\frac{1-q_2}{c_2} - q_2 \tag{4.45}$$

Thus (4.44) represents the inverse of the reaction function $q_1 = \phi_1(q_2)$, (4.45) the inverse of the reaction function $q_2 = \phi_2(q_1)$. As negative outputs make no sense, (4.44) is restricted to the interval $q_2 \in [0, \overline{q}_2]$ and (4.45) to the interval $q_1 \in [0, \overline{q}_1]$, where $\overline{q}_2, \overline{q}_1$ solve the implicit equations $\overline{q}_2 = \ln\frac{1-\overline{q}_2}{c_2}$ and $\overline{q}_1 = \ln\frac{1-\overline{q}_1}{c_1}$. Note that $0 < \overline{q}_1, \overline{q}_2 < 1$. Further, note that in these intervals of restriction (4.44) and (4.45) are one-to-one with negative slopes.

These slopes can be calculated from (4.44) and (4.45) through simple differentiation

$$\frac{dq_2}{dq_1} = -\frac{2-q_1}{1-q_1} \tag{4.46}$$

$$\frac{dq_1}{dq_2} = -\frac{2-q_2}{1-q_2} \tag{4.47}$$

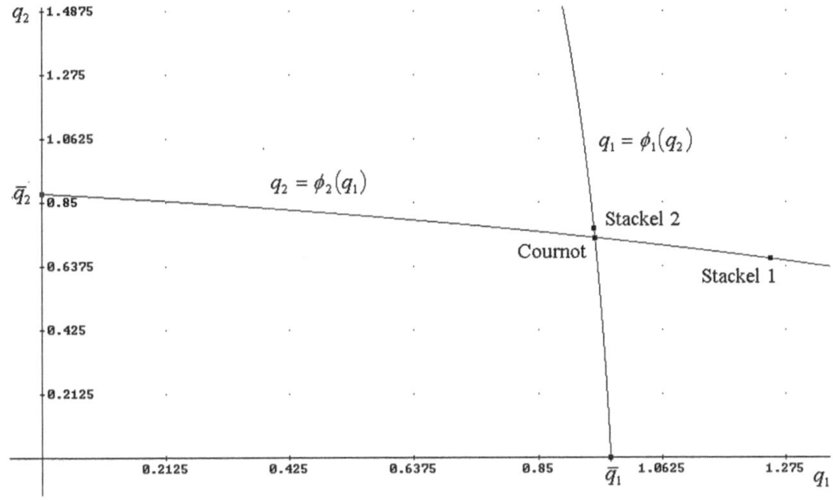

Fig. 4.8 Cournot reaction functions, the (Cournot) equilibrium at their intersection, and the two Stackelberg equilbria, (Stackel 1) with firm 1 as leader, and (Stackel 2) with firm 2 as leader. Parameters $c_1 = 0.01, c_2 = 0.05$

Again, as these are computed from the inverse, the derivatives of the reaction functions are rather $\phi_1'(q_2) = -\frac{1-q_1}{2-q_1}$ and $\phi_2'(q_1) = -\frac{1-q_2}{2-q_2}$. Like $\phi_1(q_2)$ and $\phi_2(q_1)$ are functions of q_2 and q_1 respectively, one might prefer the same for their derivatives, but this would require substitutions for the inverses of (4.44) and (4.45), and these cannot be obtained in closed form.

Anyhow one can get

$$\phi_1'(q_2) \cdot \phi_2'(q_i) = \frac{(1-q_1)(1-q_2)}{(2-q_1)(2-q_2)} < \frac{1}{4} \tag{4.48}$$

in the restriction where

$$0 < q_1 < \bar{q}_1 < 1 \quad \text{and} \quad 0 < q_2 < \bar{q}_2 < 1 \tag{4.49}$$

Hence, in any possible intersection point for the reaction functions (4.44) and (4.45), the product of their derivatives is less than one fourth, i.e., less than unity. As (4.44) and (4.45) have zero direct derivatives, the product is the Jacobian of (4.44) and (4.45) considered as a dynamical system. This implies that any intersection point of the reaction functions, i.e., any Cournot equilibrium point, is always stable.

Figure 4.8 illustrates a pair of reaction functions for the parameter combination $c_1 = 0.01$, $c_2 = 0.05$, which will be repeatedly used for exemplification below. From $\overline{q}_2 = \ln\frac{1-\overline{q}_2}{0.05}$ and $\overline{q}_1 = \ln\frac{1-\overline{q}_1}{0.01}$ one can calculate the upper bounds $\overline{q}_2 = 0.8795$ and $\overline{q}_1 = 0.9735$ for the admissible ranges of the inverse reaction functions (4.44) and (4.45).

Cournot Equilibrium

The intersection point of the reaction functions is the Cournot equilibrium. Its coordinates can be determined through solving (4.44) and (4.45) as a simultaneous system of equations for q_1 and q_2. Unfortunately, an analytical solution is not available, but for any given parameter combination a numerical solution is possible. For, $c_1 = 0.01$, $c_2 = 0.05$, illustrated in Fig. 4.8, the numerical solution is

$$q_1 = 0.94642$$
$$q_2 = 0.73211$$

This point is marked at the intersection of reaction curves in Fig. 4.8. The corresponding profits

$$\Pi_1 = 0.16718$$
$$\Pi_2 = 0.10004$$

can then be calculated from (4.40) and (4.41). Obviously the firm with lower variable unit costs (firm 1) obtains the higher profit.

Traditional Stackelberg Equilibrium

Suppose the first competitor acts as a Stackelberg leader, an assumption that will be kept throughout this digression; the facts are just the reverse for the second. The leader takes the reaction function $q_2 = \phi_2(q_1)$ for granted, substitutes it in its profit expression (4.40) and maximizes through putting the total (not just partial) derivative equal to zero.

An apparent problem is that only an expression for the inverse reaction function $q_1 = \phi_1^{-1}(q_2)$ is available in (4.45). But, never mind, the reaction function and its inverse are monotonic one-to-one in the restricted range $q_2 \in [0, \overline{q}_2]$, so one can substitute (4.45) in (4.40), and in stead equate the derivative with respect to q_2 zero. After solving, q_2 can be substituted back in (4.45) to obtain the corresponding q_1.

Hence, substituting (4.45) in (4.40), profits are

$$\Pi_1 = \left(\frac{c_1}{1-q_2} - c_1\right)\left(\ln\frac{1-q_2}{c_2} - q_2\right) \tag{4.50}$$

with the q_2 derivative

$$\frac{d\Pi_1}{dq_2} = \frac{c_2}{(1-q_2)^2}\left(\ln\frac{1-q_2}{c_2} - 2\right) + c_1\frac{2-q_2}{1-q_2} \tag{4.51}$$

Equated to zero (4.51) yields

$$\ln\frac{1-q_2}{c_2} + \frac{c_1}{c_2}(1-q_2)(2-q_2) = 2 \tag{4.52}$$

Again, it is not possible to solve (4.52) by analytical means, but, using $c_1 = 0.01$, $c_2 = 0.05$, the numerical solution is $q_2 = 0.66245$, which substituted in (4.45) yields $q_1 = 1.24725$. This point

$$q_1 = 1.24725$$
$$q_2 = 0.66245$$

is marked and labelled “Stackel 1” on the graph of the reaction function $q_2 = \phi_2(q_1)$ in Fig. 4.8. The corresponding profits become

$$\Pi_1 = 0.17227$$
$$\Pi_2 = 0.06500$$

Not surprisingly the leader gets a higher profit, the follower a lower profit than in Cournot equilibrium.

Periodic Leadership

The following question now arises: Why does the Stackelberg leader need to keep a constant supply over time? Given it is granted that the follower always adheres to its proper reaction function, the leader may devise a supply policy variable over time, and so perhaps obtain a higher leadership profit than from a constant supply. This is what will be studied next.

Start with the simplest 2-period case. It is not only the simplest, but numerical experiment indicate that the 2-period cycle is superior to any other periodicity. [1]

In a 2-period leadership cycle the leader switches between two supplies, q_1^o and q_1^e, the suffices referring to odd and even periods. The follower then responds with q_2^e and q_2^o one period later.

For the follower the inverse reaction functions are

$$q_1^o = \ln\frac{1-q_2^e}{c_2} - q_2^e \tag{4.53}$$

$$q_1^e = \ln\frac{1-q_2^o}{c_2} - q_2^o \tag{4.54}$$

---

[1] However, the case has certain drawbacks: (a) A 2-period cycle is extremely easy to identify by the follower and adapt to through incorporating a 2-period delay in the responses.(b) The cycle tends to make the follower run at an average loss over time which might make the situation untenable.

The leader assumes that the follower adheres to these reaction functions, and is now interested in the average profits over two subsequent periods

$$\Pi_1 = \tfrac{1}{2}\left(\Pi_1^o + \Pi_1^e\right) = \tfrac{1}{2}\left(\left(e^{-q_1^o - q_2^o} - c_1\right)q_1^o + \left(e^{-q_1^e - q_2^e} - c_1\right)q_1^e\right) \quad (4.55)$$

The leader maximizes (4.55) with respect to q_1^o and q_1^e, but takes in account the constraints (4.53) and (4.54). Again, there are only inverses available, but, substituting for q_1^o and q_1^e, one can maximize with respect to q_2^e and q_2^o in stead.

For dealing with the maximisation one needs two copies of (4.47), for the alternating situations

$$\frac{dq_1^o}{dq_2^e} = -\frac{2-q_2^e}{1-q_2^e} \quad (4.56)$$

$$\frac{dq_1^e}{dq_2^o} = -\frac{2-q_2^o}{1-q_2^o} \quad (4.57)$$

Differentiating (4.55) with respect to q_2^e and q_2^0 yields the two conditions,

$$\frac{\partial \Pi_1}{\partial q_2^o} = \frac{1}{2}\left(\left(1-q_1^e\right)e^{-q_1^e - q_2^e} - c_1\right)\frac{dq_1^e}{dq_2^o} - \frac{1}{2}q_1^o e^{-q_1^o - q_2^o} = 0 \quad (4.58)$$

$$\frac{\partial \Pi_1}{\partial q_2^e} = \frac{1}{2}\left(\left(1-q_1^o\right)e^{-q_1^o - q_2^o} - c_1\right)\frac{dq_1^o}{dq_2^e} - \frac{1}{2}q_1^e e^{-q_1^e - q_2^e} = 0 \quad (4.59)$$

Substituting for the derivatives from (4.56) and (4.57),

$$\frac{\partial \Pi_1}{\partial q_2^o} = -\frac{1}{2}\frac{2-q_2^o}{1-q_2^o}\left(\left(1-q_1^e\right)e^{-q_1^e-q_2^e}-c_1\right)-\frac{1}{2}q_1^o e^{-q_1^o-q_2^o}=0 \tag{4.60}$$

$$\frac{\partial \Pi_1}{\partial q_2^e} = -\frac{1}{2}\frac{2-q_2^e}{1-q_2^e}\left(\left(1-q_1^o\right)e^{-q_1^o-q_2^o}-c_1\right)-\frac{1}{2}q_1^e e^{-q_1^e-q_2^e}=0 \tag{4.61}$$

Upon substitution from (4.53) and (4.54) there remain two equations to determine q_2^0 and q_2^e, but these are too messy to be of much use analytically. Further, (4.60) and (4.61) do not provide any reasonable inner solutions except the traditional Stackelberg case where $q_1^o = q_1^e$. Figure 4.9 illustrates the reason. The profit level is depicted in terms of shading, brighter shade indicating higher profit. One can see the traditional Stackelberg equilibrium as an inner maximum on the diagonal. In that case (4.60) and (4.61) become identical.

However, the shades at the right and upper boundaries are even brighter, indicating that a higher profit can indeed be obtained through an oscillating leadership action. As the colour obviously continues to become brighter in the areas located beyond the right and upper boundaries, profits continue to increase. These are not shown because beyond the bounded area the follower's reaction becomes negative, so the solution ceases to be meaningful. The leader has to stop at the boundaries where the follower stops to produce.

One therefore has to look for *boundary* maxima, either at the right or the upper boundary. As a consequence then only *one* of the equations (4.60) and (4.61) holds. For definiteness choose the brightest spot at the upper edge of the picture. This means that the leader produces the maximum possible quantity in even periods, and as a consequence throws the follower out from the market every odd period. (Note that one could have chosen the brightest spot on the right boundary in stead, but as an alternative. Choosing this action for both periods is very suboptimal, as one concludes from the dark shade of the upper right corner of the square in Fig. 4.9.

Accordingly the leader chooses

$$q_1^e = -\ln c_2 \tag{4.62}$$

From (4.54) then

$$q_2^o = 0 \tag{4.63}$$

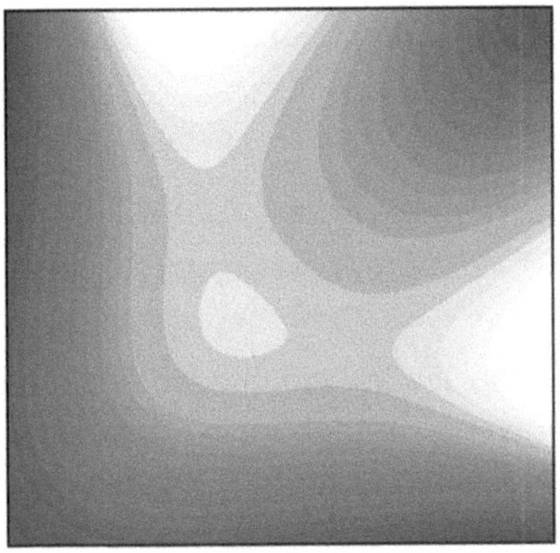

Fig. 4.9 Profit surface for two period Stackelberg leadership cycle

Due to the choice made, only (4.61), but not (4.60), remains applicable for determining leadership action in odd periods. Substituting (4.53), (4.62) and (4.63) simplify (4.61) to

$$\frac{\partial \Pi_1}{\partial q_2^e} = -\frac{1}{2}\frac{2-q_2^e}{1-q_2^e}\left(\left(1-\ln\frac{1-q_2^e}{c_2}+q_2^e\right)\frac{c_2}{1-q_2^e}e^{q_2^e}-c_1\right) \qquad (4.64)$$
$$+\frac{1}{2}c_2\ln\left(c_2 e^{-q_2^e}\right)=0$$

which determines q_2^e. From (4.53) then q_1^o is determined.

Using the numerical parameter values $c_1 = 0.01, c_2 = 0.05$, the numerical solution $q_2^e = 0.71688$ can be obtained from (4.64), which substituted, along with $c_2 = 0.05$, in (4.53) gives $q_1^o = 1.01697$. Using $c_2 = 0.05$ in (4.61), one obtains $q_1^e = 2.99573$, and, from (4.62), $q_2^o = 0$. Collecting results, the leadership 2-cycle is as follows

	o	e
q_1	1.01697	2.99573
q_2	0	0.71688

Using these values along with $c_1 = 0.01, c_2 = 0.05$ in (4.40)- (4.41), average profits are

$$\Pi_1 = 0.20042$$
$$\Pi_2 = -0.00917$$

Notably, profits for the leader are indeed higher than in the stationary Stackelberg point. However, as mentioned, there are problems with this case. The leader throws the follower out from the market every second period, so these periods the follower does not make any profit at all, but, even worse, the intervening periods the follower makes a loss which recurs every cycle and just accumulates over time.

The follower might then just leave the market. However, why should the follower drop out? The 2-period cycle is extremely simple and easy to observe. The follower could then use a two period delay in forming expectations, and hence correctly anticipate every move by the leader.

It can be shown that such learning and adaptation makes the system land in the stationary Stackelberg equilibrium. See Puu (2010). The formal argument will not be repeated here.

The leader can also try other periodicities. One that is particularly interesting is arranging a 3-period cycle. The analysis naturally becomes more messy for longer cycles, so just note that also in the 3-period cycle the leader throws the follower out from the market one of the periods in the cycle. The profit levels as shades in 3-space can be seen in Fig. 4.10. This time the Stackelberg equilibrium point cannot be seen, as it is hidden inside the block, but the three alternative boundary optima are clearly visible.

Using the same computation procedure as in the 2-period case, the numerical results for the 3-cycle, again with $c_1 = 0.01, c_2 = 0.05$, and choosing the first of every three period sequence for throwing out the competitor are

q_1	1.0969	1.0366	2.9957
q_2	0	0.6988	0.7125

Profits averaged per period then become

$$\Pi_1 = 0.1904$$
$$\Pi_2 = 0.0234$$

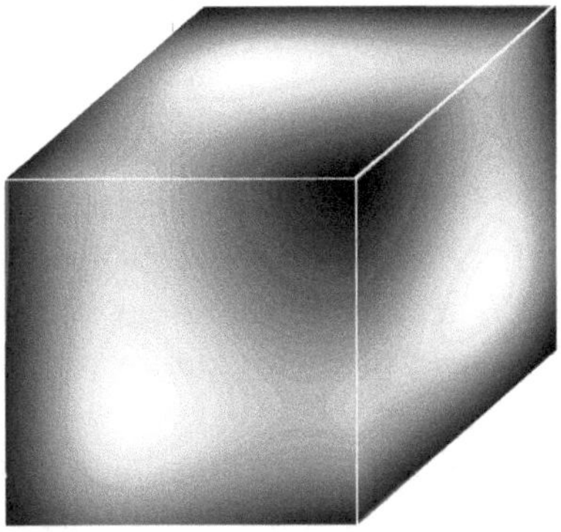

Fig. 4.10 Three period leadership profit shown in terms of shading

Obviously, in a 3-cycle the leader earns less than in the 2-cycle, but more than in Stackelberg stationary equilibrium. On the other hand, the follower still makes an average profit, even if it is small.

Checking different longer periodicities, it shows up as a general result that the leader makes the most advantageous result by arranging every even periodicity as a sequence of 2-cycles, every odd periodicity as a composition of just one 3-cycle and again a sequence of 2-cycles. See Puu (2010).

Skipping messy derivations, just one example of a 7-period cycle is given here, again with parameters $c_1 = 0.01, c_2 = 0.05$.

q_1	1.01697	2.9957	1.0969	1.0366	2.9957	1.01697	2.9957
q_2	0	0.71688	0	0.6988	0.7125	0	0.71688

This is a composition of one 2-cycle, one 3-cycle, and again one 2-cycle. The average profits then become:

$$\Pi_1 = 0.1961$$
$$\Pi_2 = 0.0048$$

Note that $0.1961 = \frac{4}{7} \cdot 0.2004 + \frac{3}{7} \cdot 0.1904$, so the resulting profits in a 7-period cycle is a weighted average of the profits that accrue for the 2-cycle and for the 3-cycle, where the weights depend on the number of periods belonging to these two types of cycles.

The remaining question concerns learning and adapting by the follower. To deal with this, consider that the 2-cycle consists of two outputs repeated again and again. But it does not matter how these are ordered; as illustrated for the 7-cycle, average profits are determined by the relative frequencies alone, and not by their order.

The sequence of 2-cycles can be considered a binary sequence. *It can be arranged in a way which Popper (1959) used for defining randomness, or as used in modern chaos literature to represent unpredictable sequences. See Devaney (1989). Obviously such random or unpredictable sequences can never be learned*, or adapted to.

The main conclusion of this digression is that it is so easy to put Stackelberg leadership in a dynamic format that it is surprising that Stackelberg never thought of it.

References

Cournot, A. (1838). *Récherces sur les principes mathématiques de la théorie des richesses.* Paris: Dunot.

Devaney, R. L. (1989). *An introduction to chaotic dynamical systems.* Redwood City, CA: Addison-Wesley.

Popper, K. R. (1959). *The logic of scientific discouvery.* London: Hutchinson.

Puu, T. (2009). Unifying Cournot and Stackeberg duopoly in one dynamical system. In C. Chiarella, G. I. Bischi, L. Gardini, (Eds.), *Nonlinear dynamics in economics and finace*, Berlin: Springer.

Puu, T. (2010). Dynamics of Stackelberg duopoly. In T. Puu, A. Panchuk. (Eds.), *Advances in nonlinear economic dynamics.* New York: Nova Science Publishers Inc.

von Stackelberg, H. (1934). *Marktform und Gleichgewicht.* Berlin: Julius Springer.

5 Capacity Limits

Stability and Capacity Limits

In Chap. 2 the Cournot–Theocharis stability issue, which will be called so, though it was raised 20 years earlier by Palander, was discussed. The problem was that in certain model families increasing competition through adding new competitors destabilised Cournot equilibrium. Accordingly, the idea of a straight path leading from monopoly, over oligopoly, to perfect competition, where profits were eliminated and marginal cost pricing was used, became subject to doubt. What interest do the properties of an equilibrium have, after all, if it becomes unstable?

It was proposed that the whole problem rested on mistaken assumptions with respect to the issue of relevance. The problematic assumptions were related to constant returns to scale in production, leading to constant marginal cost. A firm facing constant marginal cost is in principle infinitely large as it can without suffering increases in production cost increase supply to any degree, and so magnify the tiniest difference between market price and marginal cost to limitless profit just by increasing the scale of operation.

It was suggested that destabilisation through adding competitors of an infinite size was neither surprising nor interesting. What one wants to compare is situations of many small to few large competitors. However, large and small is impossible to model without introducing capacity limits.

Such could be defined as blunt limits under circumstances of constant marginal costs, which just jump to infinity when the limit is reached. However, it is also possible to choose a smooth variant, where the capacity limits are reached through asymptotically rising costs.

Fortunately, this can be modelled using one of the most popular production functions from current microeconomics literature, the constant elasticity of substitution (CES) function. See Puu (2007, 2008) and Puu and Panchuk (2009).The concept of elasticity of substitution was introduced by Sir John Hicks in 1932.

T. Puu, *Oligopoly: Old Ends – New Means*, DOI 10.1007/978-3-642-15964-0_5,

The concept of substitution elasticity is in itself a rather weird measure of the curvature of an isoquant, quite different from accepted mathematical concepts such as the radius of curvature of an osculating circle. It seems to have been Dickinson (1954), who invented this production function, assuming the elasticity of substitution to be constant on each isoquant. The isoquants then, by the way, reinvent another mathematical concept, the Minkowski metric. The CES function became popular through Arrow et al. (1961), and supplemented the Cobb–Douglas function for empirical production studies.

The use made of the CES function below, connects to another branch of production and growth studies. Ragnar Frish, the great pioneer in production studies since the 1920s (even if he was copied by others who even published his results before him), claimed that decreasing returns to scale were a matter of how many inputs were variable. If all inputs were variable, which never happened in practice, then the returns to scale would be constant. Whenever some inputs were fixed then variable returns, first increasing, eventually decreasing, would be the rule. See Frisch (1965).

Then the "putty-clay" idea emerged. In the long run all inputs would be variable, and the returns thus constant. In the short run, all inputs would be fixed, with the Leontief type of fixed proportions technology as a rule. The most elaborate studies based on this idea were presented in Johansen (1972). See also Salter (1960).

In the sequel a related idea will be suggested. Not all inputs, just capital is assumed to be fixed for some time, through investing in some specific capital equipment. This equipment puts a capacity limit on production possibilities in the short run, but in the long run, when the equipment is worn out and needs to be replaced, then, under the same production assumptions, returns are constant again, though only until a new equipment has been purchased.

The CES Production and Cost Functions

The normal form of the CES production function with two inputs, capital k and labour l is

$$q = \gamma\left(\delta k^{-\rho} + (1-\delta) l^{-\rho}\right)^{-\frac{1}{\rho}} \tag{5.1}$$

For increasing the possibility of a decent fit to statistical data it is good to have as many parameters as possible, but for theoretical purposes one would like to reduce their number. Redefine the variables $\alpha k \to k$, $\beta l \to l$, $\gamma q \to q$,

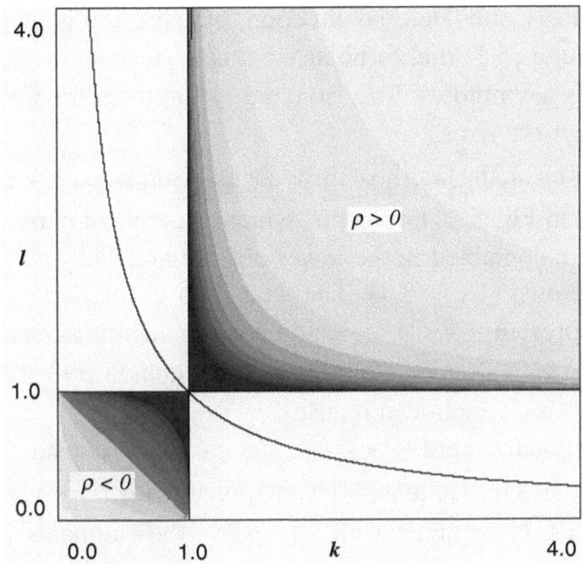

Fig. 5.1 Unit isoquants for the CES production function in the factor diagram

using $\alpha, \beta > 0$, such that $\beta / \alpha = \left((1-\delta)/\delta\right)^{\frac{1}{\rho}}$ and $\gamma / \alpha = 1/\delta^{\frac{1}{\rho}}$. Then the much nicer symmetric form

$$q = \left(k^{-\rho} + l^{-\rho}\right)^{-\frac{1}{\rho}} \tag{5.2}$$

containing just one parameter ρ emerges.

As for ρ, it is known that for $-1 < \rho < 0$, one gets the normal shapes of the isoquants (convex towards the origin), whereas for $-\infty < \rho < -1$, the convexity turns the wrong way, for instance $\rho = -2$ results in a circular arc. All cases with $\rho < 0$ have long been well known from mathematical physics as Minkowski metrics.

The case $\rho > 0$ seems to have been little utilized. All cases mentioned above are known to have isoquants with axis intercepts, whereas for $\rho > 0$ there are none. In stead the isoquants approach asymptotes. There has been some misunderstanding about these. Arrow et al. (1961) claimed that "*when $\rho < 0$, the isoquants intersect the k and l axes, while when $\rho > 0$, the isoquants only approach the axes asymptotically*". The statement is incorrect; only the Cobb–Douglas function, which often is included in the CES family though (5.1) makes no sense when $\rho = 0$, has isoquants that have the axes as asymptotes. The isoquant asymptotes for CES functions with $\rho > 0$ *lie off the axes.*

To make things clear, the family of different unit isoquants, $q = 1$, for various ρ are displayed in Fig. 5.1. For $\rho < 0$, which is the most explored case the unit isoquants are contained in the lower unit square. They all intersect the k,l-axes in the points (1,0), (0,1). The isoquants in the darker part, above the diagonal, represent $-\infty < \rho < -1$, and are considered to have the wrong kind of convexity for a production function. They include $\rho = -2$, which is the case of a circular arc (Euclidean metric).

Below the diagonal where $-1 < \rho < 0$, the cases are considered as being legitimate candidates for isoquants. The diagonal, with $\rho = -1$, is the case of perfect substitutes. Note that for all $-1 < \rho < 0$, the isoquants intersect the axes at tangency.

In the upper part of Fig. 5.1, the presently interesting cases with $\rho > 0$ can be seen. Obviously they are all alike in a topological sense, of hyperbola shape with the lines $k = 1$ and $l = 1$ as asymptotes for the unit isoquants. Again recall that the unit isoquants for $q = 1$ are considered in this picture. For higher isoquants at a fixed ρ, these asymptotes are just scaled up in proportion.

As a final simplification of the formula use the coordinate transformations $q^{\rho} \to q,\ k^{\rho} \to k,\ l^{\rho} \to l$. Given this, the CES function now reads

$$q = \frac{kl}{k + l} \tag{5.3}$$

It is easy to derive the corresponding cost functions. Denoting capital rent by r and wage rate by w, total production cost is

$$C = rk + wl \tag{5.4}$$

Solving (5.3) for $l = kq / (k - q)$, and substituting in (5.4) one obtains

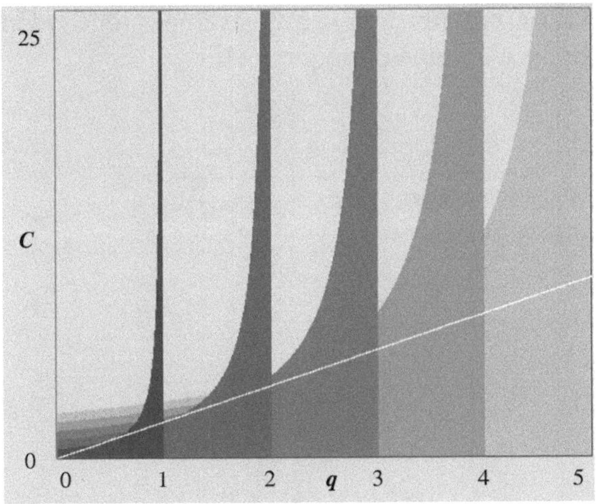

Fig. 5.2 Short and long run total cost functions

$$C = rk + w\frac{kq}{k-q} \tag{5.5}$$

which can be interpreted as a short run total cost function. If one takes capital k as fixed in the short run as suggested in the introduction, then (5.5) has a built in capacity limit because

$$\lim_{q \to k} C = \infty \tag{5.6}$$

Figure 5.2 illustrates cost functions (5.5) for different choices of fixed capital stocks. Obviously, as all these curves intersect, a small capital stock is more advantageous at small output volumes, whereas a large capital stock is preferable at large volumes.

There is a capital stock which is optimal for each production volume, which is shown by the straight line tangent to all the short run functions. It is the long run cost function when capital is fully variable (constant returns) that has this envelope property, but it cannot be accessed each and every period when the firm has already chosen a durable capital equipment on the basis of previous market conditions even if the current conditions are different. Then only the short run functions remain as possibilities.

To find the long run cost function, use its envelope property through optimising capital stock for each volume of output, putting

$$\frac{\partial C}{\partial k} = r - w\frac{q^2}{(k-q)^2} = 0 \tag{5.7}$$

Solving,

$$k = \left(1 + \sqrt{\tfrac{w}{r}}\right)q, \tag{5.8}$$

is obtained.

Substituting back from (5.8) in (5.5)

$$C = r\left(1 + \sqrt{\tfrac{w}{r}}\right)q + w\frac{\left(1 + \sqrt{\tfrac{w}{r}}\right)q^2}{\sqrt{\tfrac{w}{r}}q} = \left(\sqrt{r} + \sqrt{w}\right)^2 q \tag{5.9}$$

Defining the constant

$$c = \left(\sqrt{r} + \sqrt{w}\right)^2 \tag{5.10}$$

as a long run marginal cost, the long run cost function can be written

$$C = cq \tag{5.11}$$

The long run marginal cost is a constant. The short run marginal cost can easily be obtained from (5.5) as

$$\frac{\partial C}{\partial q} = w\frac{k^2}{(k-q)^2} \tag{5.12}$$

which, of course, varies with output. Like total cost it goes to infinity as output approaches the capacity limit, set by the existent capital stock. Note that the initial marginal cost at zero production is

$$\lim_{q \to 0} w \frac{k^2}{(k-q)^2} = w \tag{5.13}$$

The initial short run marginal cost hence equals the wage rate. Note that from (5.10) this is always lower than the long run marginal cost that also includes capital rent. The reciprocals of w and c turn up in nonnegativity constraints for the respective reaction functions. The long run constraint will thus always be stricter than the short run constraint; which is quite natural as in the long run a firm must be able to finance both fixed and variable costs, whereas in the short run the fixed costs are there anyway and it is sufficient just to cover variable costs even when earning less than needed to fully cover the capital costs.

Stability conditions later on will depend on the wage rate as compared to capital rent. Labour force is the only variable input included in the model, so such conditions have nothing to do with labour force in particular, but rather with the relation between variable and fixed costs.

Oligopoly with Capacity Limits

Again, the isoelastic inverse demand function is

$$p = \frac{1}{q_i + Q_i} \tag{5.14}$$

where

$$Q = \sum_{i=1}^{i=n} q_i \tag{5.15}$$

and

$$Q_i = Q - q_i \, . \tag{5.16}$$

This demand function will now be combined with the short run cost function (5.5). As for the long run, the facts emergent with (5.11) are already known.

Short Run Reaction Functions

Profits are now

$$\Pi_i = \frac{q_i}{q_i + Q_i} - rk_i - w\frac{k_i q_i}{k_i - q_i} \tag{5.17}$$

where the cost function was taken from (5.5), though now including an identification index for the competitors. One might also have included a time period identification, but in order to save space just recall that all the lower case variables for planned capital stock and output have $t+1$ period attribution, whereas the expected upper case residual supply has a t period attribution. As for capital, recall that in the short run process it is not changed, so it could be attributed to period t as well as to period $t+1$.

Differentiating (5.17) with respect to q_i, and equating to zero,

$$\frac{Q_i}{(q_i + Q_i)^2} = w\frac{k_i^2}{(k_i - q_i)^2} \tag{5.18}$$

is obtained. The equality is between marginal revenue on the left, which depends on the residual supply from the competitors, and marginal cost. The optimality condition (5.18) is illustrated in Fig. 5.3, with the marginal revenue curves as a family for different residual supplies, combined with the short run marginal cost curve. The intersections give the optimum points which provide the basis for the derivation of the reaction function.

The expression (5.18) can be simplified by taking square roots of both sides. Except for the three squared terms, the residual supply Q_i and the wage rate w are positive; further under the capacity constraint, $q_i < k_i$, no ambiguity can arise from the denominator of the right hand side. Hence, the linear equation

$$\frac{\sqrt{Q_i}}{q_i + Q_i} = \sqrt{w}\frac{k_i}{k_i - q_i} \tag{5.19}$$

is obtained. It is easily solved, and gives

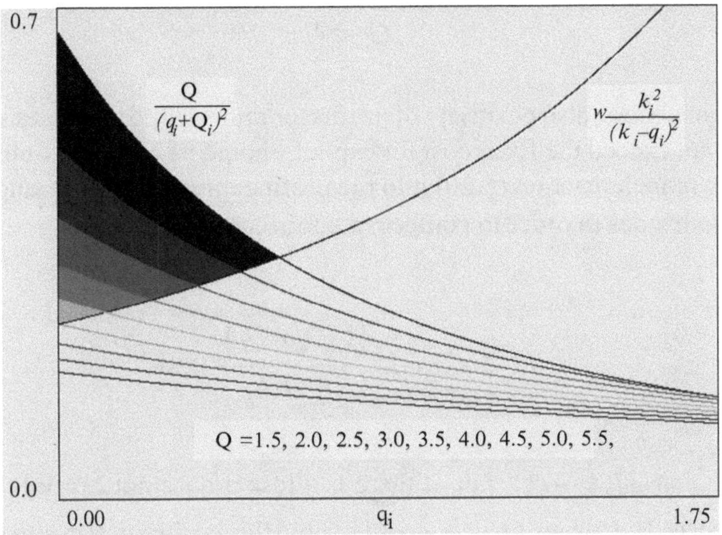

Fig. 5.3 Marginal cost curve and marginal revenue curves for various residual supplies

$$q_i = k_i \frac{\sqrt{\frac{Q_i}{w}} - Q_i}{\sqrt{\frac{Q_i}{w}} + k_i} \tag{5.20}$$

Fortunately, it was possible to find the main branch (5.20) of the reaction function even for this case. Unfortunately, however, there is no chance to calculate any closed form Cournot equilibrium coordinates. Further, the concept of Cournot equilibrium in the short run itself becomes a little ambiguous. Even if one could solve the set of simultaneous equations, in the course of time some firm would find the capital stock worn out and have to replace it, hence changing the whole system. It seems that only the long run process with capital included among the variables, which reverts the production conditions to constant returns, would be suitable for defining a Cournot equilibrium. Fortunately, for that case it is easy to obtain the equilibrium coordinates.

It is now time to consider that the main branch of the dynamic reaction function (5.20) results in a positive supply if and only if

$$Q_i < \tfrac{1}{w}. \tag{5.21}$$

For completeness, also positivity of profits, or rather a limit to losses such that they do not exceed the fixed cost for capital, should be checked. Substituting the main branch reaction (5.20) into the profit expression (5.17), and adding rk_i to both sides in order to compensate for fixed costs, yields

$$\Pi_i + rk_i = k_i \frac{\left(1-\sqrt{wQ_i}\right)^2}{Q_i + k_i} > 0, \tag{5.22}$$

showing that $\Pi_i > -rk_i$, i.e., if there is a loss it does not exceed the fixed capital costs. But, when $Q_i < \frac{1}{w}$ does not hold this positivity is due to negative costs exceeding negative revenues. Anyhow there is no need for more than (5.21). The reaction function should thus be reformulated as

$$q_i(t+1) = \begin{cases} k_i(t)\dfrac{\sqrt{\frac{Q_i(t)}{w}} - Q_i(t)}{\sqrt{\frac{Q_i(t)}{w}} + k_i(t)} & Q_i(t) < \frac{1}{w} \\ 0 & Q_i(t) \geq \frac{1}{w} \end{cases} \tag{5.23}$$

where period attributions have now been added for the dynamic process. In this statement the fact was used that in the short run process capital stock is not changed, so that $k_i(t+1) = k_i(t)$.

For the study of stability, the derivative of the main branch of the reaction function (5.20) is recorded

$$\frac{\partial q_i}{\partial Q_i} = -k_i \frac{(Q_i - k_i)\sqrt{\frac{Q_i}{w}} + 2k_i Q_i}{2Q_i\left(k_i + \sqrt{\frac{Q_i}{w}}\right)^2} \tag{5.24}$$

It looks quite messy, but will only be needed for the Cournot equilibrium point, for which case the expression is considerably simplified.

Long Run Reaction Functions

The long run case is known from Chap. 3, so the derivations can now be much shorter. The profit function is

$$\Pi_i = \frac{q_i}{q_i + Q_i} - c q_i \,. \tag{5.25}$$

Upon differentiation the optimality condition

$$\frac{Q_i}{(q_i + Q_i)^2} = c \tag{5.26}$$

is obtained. Again, it is possible to take square roots of both sides, and get the linear equation

$$\frac{\sqrt{Q_i}}{q_i + Q_i} = \sqrt{c} \tag{5.27}$$

which is readily solved for the main branch of the reaction function

$$q_i = \sqrt{\frac{Q_i}{c}} - Q_i \,. \tag{5.28}$$

This can only be used if

$$Q_i < \tfrac{1}{c} \tag{5.29}$$

As $c < w$, the condition is stricter than (5.21). Again, check profits. Substituting from (5.28) in (5.25)

$$\Pi_i = \frac{\sqrt{\frac{Q_i}{c}} - Q_i}{Q_i} - c\left(\sqrt{\frac{Q_i}{c}} - Q_i\right) = \left(1 - \sqrt{cQ_i}\right)^2 > 0 \tag{5.30}$$

is obtained, but, again, when (5.29) is not fulfilled, then positivity of profits is attained trough negative costs outweighing negative revenues, which makes no factual sense.

The long run map can now be formulated

$$q_i(t+1) = \begin{cases} \sqrt{\dfrac{Q_i(t)}{c}} - Q_i(t) & Q_i(t) < \dfrac{1}{c} \\ 0 & Q_i(t) \geq \dfrac{1}{c} \end{cases} \tag{5.31}$$

Finally, record the derivative of the main branch (5.28) of the reaction function

$$\frac{\partial q_i}{\partial Q_i} = \frac{1}{2}\frac{1}{\sqrt{cQ_i}} - 1 \tag{5.32}$$

It can be recognized from Chap. 3.

The Cournot Equilibrium

From the long run optimality conditions (5.26) and the definitions (5.15) and (5.16) it is now possible to derive the coordinates for the Cournot equilibrium point. Multiplying through (5.26) by the left hand side denominator, and using definition (5.16),

$$Q_i = c\left(q_i + Q_i\right)^2 = cQ^2 . \tag{5.33}$$

As the right hand sides are constant, the sum over index i is just

$$\sum_{i=1}^{i=n} Q_i = ncQ^2 . \tag{5.34}$$

On the other hand this sum can be derived directly from (5.16),

$$\sum_{i=1}^{i=n} Q_i = \sum_{i=1}^{i=n} (Q - q_i) = (n-1)Q \tag{5.35}$$

Next, identify the right hand sides of (5.34) and (5.35), cancel one power Q, which is nonzero, and solve for total supply

$$Q = \frac{n-1}{n}\frac{1}{c}\,. \tag{5.36}$$

Once total supply is known, residual supply

$$Q_i = \left(\frac{n-1}{n}\right)^2 \frac{1}{c} \tag{5.37}$$

is obtained from (5.32), and the individual supplies from definition (5.15)

$$q_i = \frac{n-1}{n^2}\frac{1}{c} \tag{5.38}$$

Finally, use the optimality condition for capital (5.8) to obtain

$$k_i = \left(1 + \sqrt{\tfrac{w}{r}}\right)\frac{n-1}{n^2}\frac{1}{c} \tag{5.39}$$

The Cournot point coordinates can now be used in (5.24) and (5.32) to simplify the derivatives of the reaction functions which will be needed for the study of stability. Thus, from (5.24), the reaction functions for firms that do not renew capital become

$$\frac{\partial q_i}{\partial Q_i} = -\frac{1}{2}\frac{n-2}{(n-1)\left(\sqrt{\tfrac{r}{w}}n+1\right)} \tag{5.40}$$

and for those that do

$$\frac{\partial q_i}{\partial Q_i} = -\frac{1}{2}\frac{n-2}{n-1} \tag{5.41}$$

Now all pieces are there for discussing the stability of Cournot equilibrium, i.e., the issue opened by Palander (1936, 1939) and Theocharis (1959), but considering that all firms do not always produce under constant returns, but rather under capacity constraints, hence removing the absurd assumption that firms added to the market are all of infinite size.

Stability

Now all pieces are in place for discussing stability of the Cournot equilibrium point. It is convenient to start with the simplest case, which is long run stability. Then all competitors are reinvesting in new capital equipment, and hence have access to a constant returns production technology.

Stability in the Long Run

In the Cournot equilibrium point the derivatives of all reaction functions are as stated in (5.41). With n competitors the Jacobian matrix then is

$$\mathbf{A} = \begin{bmatrix} 0 & \cdots & -\frac{1}{2}\frac{n-2}{n-1} \\ \vdots & \ddots & \vdots \\ -\frac{1}{2}\frac{n-2}{n-1} & \cdots & 0 \end{bmatrix} \tag{5.42}$$

Expressions (5.41) enter all the off-diagonal elements, whereas the diagonal elements are zero because each competitor's reaction is influenced in the same way by the previous moves of all competitors, but not at all by its own previous move.

Accordingly, the characteristic equation has eigenvalues

$$\lambda_{1,\ldots n-1} = \frac{n-2}{n-1}, \quad \lambda_n = -\frac{1}{2}(n-2) \tag{5.43}$$

Obviously the first $n-1$ eigenvalues are well located in the unit interval, whereas the last falls outside whenever $n > 4$. This is in perfect agreement with the conclusions by Agiza (1998) and Ahmed et al. (1998).

Stability in the Short Run

The conclusions are not as unambiguous for the short run case, when all competitors have fixed capital equipment, and produce with capacity constraints set through earlier investment decisions. The derivatives of the reaction functions which enter the off-diagonal elements of the Jacobian matrix, then are as stated in (5.40). Hence

$$\mathbf{B} = \begin{bmatrix} 0 & \cdots & -\frac{1}{2}\frac{n-2}{(n-1)\left(\sqrt{\frac{r}{w}}n+1\right)} \\ \vdots & \ddots & \vdots \\ -\frac{1}{2}\frac{n-2}{(n-1)\left(\sqrt{\frac{r}{w}}n+1\right)} & \cdots & 0 \end{bmatrix} \tag{5.44}$$

is the Jacobian, and the eigenvalues are

$$\lambda_{1,\ldots n-1} = \frac{1}{2}\frac{n-2}{(n-1)\left(\sqrt{\frac{r}{w}}n+1\right)}, \quad \lambda_n = -\frac{1}{2}\frac{n-2}{\sqrt{\frac{r}{w}}n+1} \tag{5.45}$$

Again, there is no problem with the multiple eigenvalues; any stability problem is caused by the last. Whenever $n < 4$, it is obvious that $-1 < \lambda_n < 0$. If, on the contrary, $n \geq 4$ then all depends on the square root of the ratio of input costs. Provided $\sqrt{\frac{r}{w}} \geq \frac{1}{2}$ holds, then $-1 < \lambda_n < 0$ is fulfilled for all $n \geq 4$. In the contrary case, when $\sqrt{\frac{r}{w}} < \frac{1}{2}$ holds, then one can always find a sufficiently large n such that $\lambda_n < -1$.

In plain words, if the wage rate exceeds four times capital rent, then there is always a number of competitors, though possibly larger than 4, the critical number for destabilisation of the long run equilibrium, such that even the short run dynamic destabilises Cournot equilibrium. If the wage rate is less than four times capital rent, then the short run dynamic is always stable around the Cournot point.

It is necessary to add some clarification to this conclusion. First, the exact numerical values are due to the simple symmetric form of the CES function chosen for simplicity of further manipulation. The conclusion hence means that if the wage rate is not too high in comparison to the rent of capital, then the short run dynamic is always stable no matter how many the competitors are. If the wage rate exceeds some critical ratio to capital rent, then there exists some critical number of competitors such that if it is exceeded even the short run dynamic is destabilised.

Second, it was already said, but is worth repeating; labour is the only variable input in the model, capital the only fixed input. The conclusion hence does not single out labour force in particular, but should be interpreted as pertinent to the ratio of the unit cost of variable to the unit cost of fixed factors of production.

Stability of a Mixed Process

The most general case is when some firms use the long run full flexibility, whereas other firms act under capacity constraints. Say that the m first firms out of a total on n have such constraints, whereas the last $n - m$ have full constant returns flexibility. The question now is; Under which circumstances is the Cournot equilibrium then stable?

There is now a mixture of reaction function derivatives according to (5.40) and (5.41) in the Jacobian matrix. To define it succinctly first define the two m by m sub-matrices

$$\mathbf{C}_{11} = \begin{bmatrix} 0 & \cdots & -\frac{1}{2}\frac{n-2}{(n-1)\left(\sqrt{\frac{r}{w}}n+1\right)} \\ \vdots & \ddots & \vdots \\ -\frac{1}{2}\frac{n-2}{(n-1)\left(\sqrt{\frac{r}{w}}n+1\right)} & \cdots & 0 \end{bmatrix} \tag{5.46}$$

$$\mathbf{C}_{12} = \begin{bmatrix} -\frac{1}{2}\frac{n-2}{(n-1)\left(\sqrt{\frac{r}{w}}n+1\right)} & \cdots & -\frac{1}{2}\frac{n-2}{(n-1)\left(\sqrt{\frac{r}{w}}n+1\right)} \\ \vdots & \ddots & \vdots \\ -\frac{1}{2}\frac{n-2}{(n-1)\left(\sqrt{\frac{r}{w}}n+1\right)} & \cdots & -\frac{1}{2}\frac{n-2}{(n-1)\left(\sqrt{\frac{r}{w}}n+1\right)} \end{bmatrix} \tag{5.47}$$

the first with zero diagonal elements, the second with all elements filled out by the short run derivatives. Further, define the corresponding $n - m$ by $n - m$ sub-matrices based on the long run derivatives

$$\mathbf{C}_{21} = \begin{bmatrix} -\frac{1}{2}\frac{n-2}{n-1} & \cdots & -\frac{1}{2}\frac{n-2}{n-1} \\ \vdots & \ddots & \vdots \\ -\frac{1}{2}\frac{n-2}{n-1} & \cdots & -\frac{1}{2}\frac{n-2}{n-1} \end{bmatrix} \tag{5.48}$$

$$\mathbf{C}_{22} = \begin{bmatrix} 0 & \cdots & -\frac{1}{2}\frac{n-2}{n-1} \\ \vdots & \ddots & \vdots \\ -\frac{1}{2}\frac{n-2}{n-1} & \cdots & 0 \end{bmatrix} \tag{5.49}$$

It is now possible to compose the Jacobian for the mixed system

$$\mathbf{C} = \begin{bmatrix} \mathbf{C}_{11} & \mathbf{C}_{12} \\ \mathbf{C}_{21} & \mathbf{C}_{22} \end{bmatrix} \tag{5.50}$$

This may seem messy, but allows one to draw fairly precise conclusions concerning stability of Cournot equilibrium in the mixed system. The eigenvalues provide two sets of $m-1$ and $n-m-1$ multiple eigenvalues

$$\lambda_{1,\ldots m-1} = \frac{1}{2} \frac{n-2}{(n-1)\left(\sqrt{\frac{r}{w}}n+1\right)} \tag{5.51}$$

$$\lambda_{m,\ldots n-2} = \frac{1}{2} \frac{n-2}{n-1} \tag{5.52}$$

which, due to previous discourses, are within the unit interval, and therefore do not contribute to any instability.

Stability problems arise from the two last eigenvalues, which solve the characteristic equation

$$\lambda^2 + a\lambda + b = 0 \tag{5.53}$$

where

$$a = \frac{1}{2} \frac{(n-2)\left(\sqrt{\frac{r}{w}}n(m-1)+n-2\right)}{(n-1)\left(\sqrt{\frac{r}{w}}n+1\right)} \tag{5.54}$$

and

$$b = -\frac{1}{4} \frac{(n-2)^2}{(n-1)\left(\sqrt{\frac{r}{w}}n+1\right)} \tag{5.55}$$

It is well known that the eigenvalues of (5.53) are in the unit interval if the three conditions $b < 0$, $a + b > -1$, and $a - b < 1$ are fulfilled. As for $b < 0$, there is no doubt from (5.55). As soon as $n > 1$, i.e., the market is past monopoly, the condition is fulfilled.

As for the $a + b > -1$, calculate

$$a+b=\frac{2\sqrt{\frac{r}{w}}n(n-2)(n-m-1)+(n-2)^2}{4(n-1)\left(\sqrt{\frac{r}{w}}n+1\right)} \tag{5.56}$$

The expression is nonnegative (and hence exceeds -1) whenever $n \geq 2$ and $n-m \geq 1$, i.e., whenever the market is at least a duopoly, and there is at least one firm in long run adjustment. Note that if the last condition is not fulfilled then the issue is short run stability as already dealt with.

Remains the third condition $a - b < 1$, which needs a bit more discussion. Calculate

$$a-b=\frac{(n-2)\left(2\sqrt{\frac{r}{w}}n(n-m-1)+3n-6\right)}{4(n-1)\left(\sqrt{\frac{r}{w}}n+1\right)} \tag{5.57}$$

Note that for $n > 1$, the denominator is positive. Hence, $a - b < 1$ holds whenever

$$(n-2)\left(2\sqrt{\tfrac{r}{w}}n(n-m-1)+3n-6\right)<4(n-1)\left(\sqrt{\tfrac{r}{w}}n+1\right) \tag{5.58}$$

Now, rearrange, collecting all terms containing the square root on the left, and the remaining terms on the right. Note that the right hand side factorizes, and finally divide both sides by $2n$. Then, arrange the factor that multiplies the square root on the left in a convenient manner, obtaining.

$$\sqrt{\tfrac{r}{w}}\left(n(n-m-3)-2(n-m-2)\right)<-\frac{(n-4)(3n-4)}{2n} \tag{5.59}$$

There are now three different cases to check. Obviously, whenever $n > 4$, the case for long run instability, the right hand side of (5.59) is negative.

(a) At least three firms operate under constant returns. As the square root of the ratio of input prices is necessarily positive, (5.59) states that the

parenthesis multiplying this square root must be negative. Suppose $n-m \geq 3$. Then

$$n(n-m-3)-2(n-m-2) \geq n(n-m-3) \geq 0, \tag{5.60}$$

which contradicts that the left hand side of (5.59) is negative, i.e., the condition for stability.

The conclusion is that whenever there are more than four competitors, and at least three of them react in the long run manner, under constant returns, then Cournot equilibrium is unstable.

There is hence only a chance for stability when at most one or two firms react in the long run way, so it is convenient to investigate these cases separately.

(b) One firm operates under constant returns. Put $n-m = 1$ in the inequality (5.59). Then the parenthesis on the left reduces to $-2(n-1)$. Dividing (5.59) by a negative factor one has to reverse the inequality sign. Hence,

$$\sqrt{\frac{r}{w}} > \frac{(n-4)(3n-4)}{4n(n-1)} \tag{5.61}$$

is the condition for stability. Note that the right hand side of (5.61) is a uniformly increasing function of n, which is bounded by a supremum of $\frac{3}{4}$. Hence with only one firm operating under constant returns Cournot equilibrium is stable if

$$\sqrt{\frac{r}{w}} > \frac{3}{4}. \tag{5.62}$$

This is a condition of the same kind as turned up for stability when all firms operated under capacity limits, though the right hand side was then $\frac{1}{2}$. Inequality (5.62) is a bit stricter in limiting the wage rate in relation to capital rent. Remains to check the case $n - m = 2$.

(c) Two firms operate under constant returns. Substituting $n - m = 2$ in equation (5.59), makes the parenthesis on the left equal to $-n$. Again, dividing by this factor the inequality sign must be reversed and the stability condition becomes

$$\sqrt{\frac{r}{w}} > \frac{(n-4)(3n-4)}{2n^2} \tag{5.63}$$

Again the right hand side of (5.63) is uniformly increasing with n and has a supremum of $\frac{3}{2}$. Hence

$$\sqrt{\frac{r}{w}} > \frac{3}{2} \tag{5.64}$$

guarantees stability. Note that, as we dealt with the suprema in the two last cases, inequalities (5.62) and (5.64) may be unnecessarily strong when there are a finite number of competitors.

Anyhow, the ratio of capital rent to wage rate plays a crucial role for stability of the Cournot equilibrium. The model with an isoelastic demand function presents stability problems when the number of competitors exceeds four. Then the case where all competitors produce under constant returns is always unstable. If, on the contrary, most competitors have capacity constraints, stability of Cournot equilibrium depends on how many of them in a certain period are reinvesting, hence having access to constant returns technology. If their number is three or more, then no input price ratio can guarantee stability. If the number is two, then (5.64) gives a sufficient condition for stability, if the number is one, then (5.62) gives the same. Finally, as was shown earlier, if their number is zero, i.e., all firms operate under capacity constraints, then

$$\sqrt{\frac{r}{w}} > \frac{1}{2} \tag{5.65}$$

is the condition for stability. If it is not fulfilled, then there always exists a number of competitors such that even the case with all firms operating under capacity constraints is destabilised

Investment

The study of stability of Cournot equilibrium was based on some mixture of the maps (5.23) and (5.31) for firms operating under capacity constraints due to past investment decisions, and those renewing capital, or entering the market, who have access to constant returns technology.

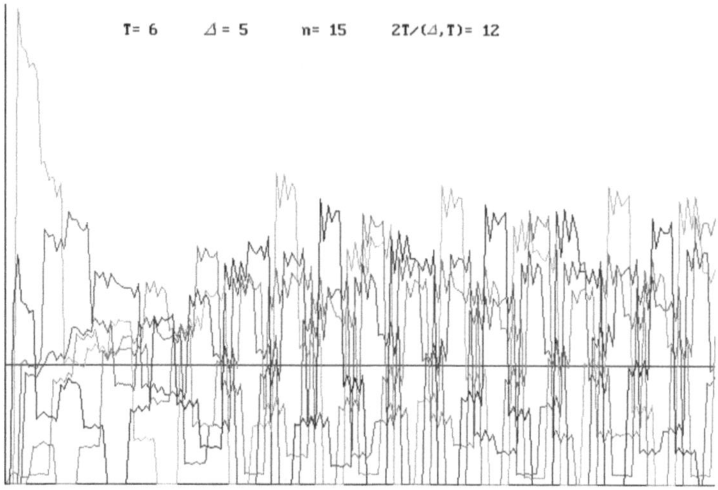

Fig. 5.4 Non-convergent time series

However, it was never stated when and on which bases the investment decisions are taken. To illustrate some of the stability issues some numerics will now be added, but then some assumption is needed for how capital renewal decisions are taken.

To this end assume a very simple process. New firms enter the industry at given intervals, whereas capital has a fixed duration while it renders full service, and then suffers sudden death. It then has to be renewed. Even if input prices are taken as fixed over time, market conditions change due to the entry of new firms. It is hence to be assumed that a firm taking a reinvestment decision will go for a different capital stock, i.e., capacity limit at each new investment decision.

Different assumptions concerning capital depreciation are imaginable, but sudden death and fixed lifetime is the simplest to start from. One might also want to internalize the reinvestment decision in the model, so that it depends on what is going on as the process evolves. Now this is taken as exogenous, but it is at least a start.

Thus take the durability of each piece of new capital to be T periods. Further, assume the firms first enter with intervals of Δ time periods, and, finally assume there are a total of n competitors. Given the fixed durations, reinvestment occurs for the firms already in the market in the same order and with the same periodicity as when they first entered the market. It is convenient to define an indicator function

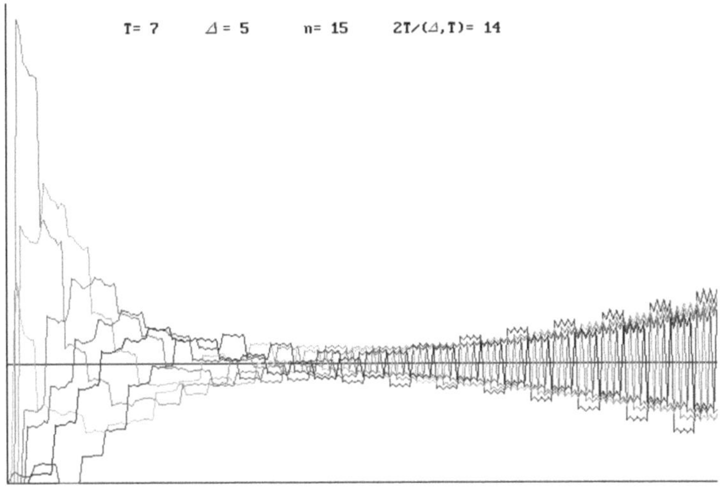

Fig. 5.5 Almost convergent time series

$$\delta(i,t) = (t - \Delta \cdot i) \bmod T \tag{5.66}$$

Suppose firm number i (re)invests in time period t when $\delta(i,t) = 0$, and does not when $\delta(i,t) \neq 0$.

Then the map (5.23) is applied whenever $\delta(i,t) \neq 0$ holds. There is also needed the supplementary assumption

$$k_i(t+1) = k_i(t). \tag{5.67}$$

If $\delta(i,t) = 0$ holds, then (5.31) is applied, with the supplementary rule stated in (5.8)

$$k_i(t+1) = \left(1 + \sqrt{\tfrac{w}{r}}\right) q(t+1) \tag{5.68}$$

To link the discussion to the stability issue discussed in the preceding section, note that, given the rule (5.66), no more than one firm reinvests in the same period if

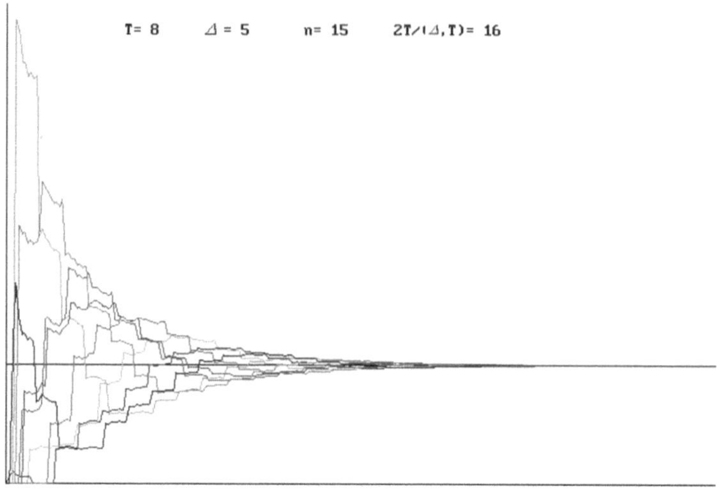

Fig. 5.6 Time series convergent to Cournot equilibrium

$$n \le \frac{T}{(\Delta, T)}$$

and never more than two firms reinvest in the same period, if

$$n \le \frac{2T}{(\Delta, T)}$$

holds. The symbol (Δ, T) denotes the greatest common divisor of these two bracketed integers. If they have none, then the value is, of course unity.

Figures 5.4–5.6 Illustrate the result of numerical experiment with the mixed model as described. In each picture the parameters and the calculated critical value for the number of competitors are stated. As one can see, this series representing an ascending series of capital durabilities goes from clear instability, over near stability, to stability of the Cournot equilibrium.

Notably much more is to be done in terms of numerical experiment. Some more results can be found in Panchuk and Puu (2009). The pictures just illustrate the stability of the Cournot point, already discussed analytically. The numerics says nothing about multistability nor about other types of attractors.

References

Agiza, H. N. (1998). Explicit stability zones for Cournot games with 3 and 4 competitors. *Chaos, Solitons & Fractals, 9*, 1955–1966.

Ahmed, E., & Agiza, H. N. (1998). Dynamics of a Cournot game with n competitors. *Chaos, Solitons & Fractals, 9*, 1513–1517.

Arrow, K. J., Cheney, B., Minhas, B., & Solow, R. M. (1961). Capital-labor substitution and economic efficiency. *Review of Economics and Statistics, 43*, 225–250.

Dickinson, H. D. (1954). A note on dynamic economics. *Review of Economic Studies*, *22*, 69–79.

Frisch, R. (1965). *Theory of production*. Dordrecht: D. Reidel Publ. Comp.

Hicks, J. R. (1932). *The theory of wages*. London: Macmillan.

Johansen, L. (1972). *Production Functions*. Amsterdam: North-Holland.

Palander, T. F. (1936). Instability in competition between two sellers, Abstracts of papers presented at the research conference on economics and statistics held by the Cowles Commission at Colorado College, *Colorado College Publications*, General Series No. 208, Studies Series No. 21.

Palander, T. F. (1939). Konkurrens och marknadsjämvikt vid duopol och oligopol. *Ekonomisk Tidskrift, 41*, 124–145, 222–250.

Panchuk, A., & Puu, T. (2009). Stability of Cournot oligopoly when the competitors are numerous. *Computers and Mathematics with Applications, 58*, 2022–2034

Puu, T. (2007). Layout of a new industry: From oligopoly to competition. *Pure Mathematics and Applications, 16*, 475–492.

Puu, T. (2008). On the stability of cournot equilibrium when the number of competitors increases. *Journal of Economic Behavior and Organization, 66*, 445–456.

Puu, T., & Panchuk, A. (2009). "Oligopoly and stability". *Chaos, Solitons, & Fractals, 41*, 2505–2516.

Salter, W. E. G. (1960). *Productivity and technical change*. Cambridge: Cambridge University Press.

Theocharis, R. D. (1959). On the stability of the Cournot solution on the oligopoly problem. *Review of economic Studies, 27*, 133–134.

6 Multistability

Historical Background

In a short printed abstract of contributions to a Cowles Commission conference at Colorado Springs in 1936, and an extensive follow up, in Swedish, dating from 1939, Tord Palander focused some interesting dynamics problems in Cournot duopoly when the demand curve was kinked linear, of the type suggested by Joan Robinson in 1933, where the marginal revenue curve jumped up, producing two different local profit maximising intersections with the curve of marginal cost. Palander took it constant or even zero.

The Cournot reaction functions, as well, then became piecewise linear, including a jump. In the duopoly market two such reaction functions could produce several, coexistent, locally stable equilibria. Palander specified the basins for these, though nor completely, and also the basin from which the system in stead went to an attractive 2-periodic oscillation.

For another case, where the reaction functions did not intersect at all, and there hence did not exist any Cournot equilibrium, Palander recognised the existence of an attractive 3-period cycle. Given the numerical tools of the period, the accuracy of his calculations is in fact amazing, though he missed the coexistence with a 6-period cycle in this latter case, and again the complete characterisation of the basins.

Palander gave his argument in terms of two numerical cases, one with identical firms, another with a slight asymmetry between them.

It is noteworthy that, also in 1936, Abraham Wald considered the same type of problems, in an article which later became celebrated as the first rigorous statement of existence problems for multi-market equilibria, left open by Walras and finally elaborated by Arrow and Debreu (1954).

Like Palander, Wald gave his argument in terms of two numerical examples, also assuming the demand curve to start and end with straight line segments. Unlike Palander, he did not assume the curves to meet under a sharp

T. Puu, *Oligopoly: Old Ends – New Means*, DOI 10.1007/978-3-642-15964-0_6,

angle. Rather the two line segments were smoothly joined by either a circular segment, or by an inverse demand curve where price was reciprocal to supply squared. The lines and curves were hence joined at tangency, so the demand function was not only continuous but even differentiable. The first example, with a circular segment, seems miscalculated, but the second, hyperbola like, is most interesting, as it results in a nondenumerable infinity of coexistent Cournot equilibria.

Wald was not interested in the dynamics produced by the reaction functions. Had he been, he would have found that none of the coexistent equilibria was even locally attractive. Rather the whole state space, except a sparse line segment, provides basins for oscillatory motion, quite as in Palander's cases, though all different and nondenumerably infinite in number.

Of these Palander's contribution is the less construed. Sectional linearity of the demand function, combined with constant marginal costs, makes all pieces of the reaction function have slope $-1/2$, and this makes all Cournot equilibria stable. There may be cycles, even coexistent, but no such things as chaos.

However, in a more general setting, falling marginal costs make the slope of the reaction functions steeper, so possibly destabilising the Cournot equilibria and resulting in much more intriguing dynamics. Globally decreasing marginal costs, of course, do not make sense, but locally, as an approximation to the falling section of a U-shaped marginal cost curve, they indeed do.

As for Palander's assumed demand function, demand sensitivity to price goes up as price goes down. Joan Robinson in 1933 gave factual arguments for this, in terms of new and more numerous groups of consumers entering the market once they could afford the good, transforming it from an exclusive luxury to something everyone consumes. This idea is used in all textbook discussions of price discrimination.

The so called "kinked demand" case for duopoly is something quite different, and not discussed here as the idea is rather trivial. Then demand is assumed to become less, not more, elastic as price goes down. The explanation for this is not in terms of the behaviour of the consumers. Rather, the competitors are assumed to retaliate to increased demand, aiming at bringing price down, but not to reduced demand to bring price up. This model was used to the purpose of explaining an allegedly observed stability in duopoly pricing. See Sweezy (1939). Here the jump in marginal revenue is down, not up, but this is not the issue Palander dealt with, which no doubt is the more interesting.

Palander's First Case

Demand and Cost Functions

In his first case Palander assumed two identical duopoly firms both having *zero marginal cost*, facing the kinked inverse demand function:

$$p = f(Q) = \begin{cases} 100 - \dfrac{Q}{20} & Q < 1{,}800 \\ \dfrac{100}{7} - \dfrac{Q}{420} & Q \geq 1{,}800 \end{cases} \tag{6.1}$$

where $Q = q_1 + q_2$. In the case of monopoly, marginal revenue is

$$MR = \frac{d(pQ)}{dQ} = \begin{cases} 100 - \dfrac{Q}{10} & Q < 1{,}800 \\ \dfrac{100}{7} - \dfrac{Q}{210} & Q \geq 1{,}800 \end{cases} \tag{6.2}$$

The Robinson type demand curve, as well as the marginal monopoly revenue curves for the identical duopolists, are shown in Fig. 6.1. Even if the picture is drawn for duopolists, they become monopolists when the competitor does not supply anything.

The picture shows a family of different marginal revenue curves for different supplies from the competitor. Those for the monopoly case are shown on top of the picture. As usual, they have the same price axis intercept as the demand function, and doubled slope as compared to that of the inverse demand function. There is a jump up from one linear piece to another, in the monopoly case from –80 to 40/7 at the kink point, i.e. for q_1 or $q_2 = 1{,}800$.

With zero marginal cost one obtains *two local profit maxima* for the monopolist from the condition $MR_i = 0, q_i = 1{,}000$ and $q_i = 3{,}000$ respectively. The corresponding monopoly prices according to (6.1) then become $p = 50$, and $p = 100/21$; the profits, given zero variable production costs, $pq_i = 50{,}000$ and $pq_i = 150{,}000/7$ respectively, so it is obvious that the monopolists would select the first of the local profit maxima.

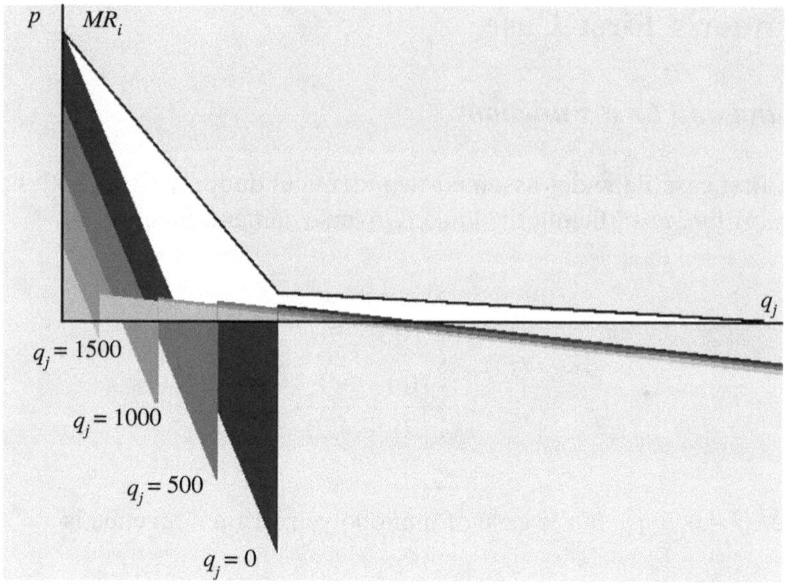

Fig. 6.1 Inverse market demand curve and marginal revenue curves for each of the identical competitors given four different supplies by the other one

The marginal revenue curve for the monopoly case is shown in Fig. 6.1 as the curve supported by the darkest shade area. As marginal costs are zero, the horizontal axis is also the marginal cost curve. It intersects the discontinuous marginal revenue curve in two points, the local profit maxima.

Economists are used to establish the global profit maximum by comparing the loss and gain areas bounded by the marginal revenue curve above and below the marginal cost curve, presently the zero line. It is obvious that in the monopoly case the loss area, below the axis, outweighs the gain area, above the axis, so that the leftmost intersection with the axis establishes the global optimum, quite as already demonstrated numerically.

This is not always the case in duopoly when there are positive supplies by the competitor. The marginal revenue curves with support areas in brighter shade have increased gain areas and decreased loss areas, so eventually the global optimum switches to the rightmost intersection.

Reaction Functions

With nonzero supply from the competitors, the marginal revenues become

$$MR_1 = \begin{cases} 100 - \dfrac{1}{10}q_1 - \dfrac{1}{20}q_2 & q_1 < 1{,}800 - q_2 \\ \dfrac{100}{7} - \dfrac{1}{210}q_1 - \dfrac{1}{420}q_2 & q_1 \geq 1{,}800 - q_2 \end{cases} \tag{6.3}$$

$$MR_2 = \begin{cases} 100 - \dfrac{1}{20}q_1 - \dfrac{1}{10}q_2 & q_2 < 1{,}800 - q_1 \\ \dfrac{100}{7} - \dfrac{1}{420}q_1 - \dfrac{1}{210}q_2 & q_2 \geq 1{,}800 - q_1 \end{cases} \tag{6.4}$$

Obviously, a positive supply by the competitor subtracts from the intercept of the marginal revenue curve, and also pushes the kink point to the left, as one sees in Fig. 6.1. Note that in equations (6.3) and (6.4) the competitor's supply enters with only half the coefficient of the duopolist's own supply. From Fig. 6.1, a positive supply from the competitor also decreases the size of the dip, thereby changing the loss and gain areas. As already noted it may be that the global optimum thus shifts from the leftmost intersection point to the rightmost as the competitor increases its supply.

Given zero marginal costs, the firms equate (6.3) and (6.4) to zero and solve the equations with respect to q_1 and q_2 respectively, thus obtaining two solutions each,

$$q_1(t+1) = \begin{cases} 1{,}000 - \dfrac{1}{2}q_2(t) \\ 3{,}000 - \dfrac{1}{2}q_2(t) \end{cases} \tag{6.5}$$

$$q_2(t+1) = \begin{cases} 1{,}000 - \dfrac{1}{2}q_1(t) \\ 3{,}000 - \dfrac{1}{2}q_1(t) \end{cases} \tag{6.6}$$

where the variables have been dated in view of the dynamic process. Given the firms are identical, the reaction functions have the same form. To choose the correct branch of (6.5) and (6.6), different considerations must be taken in account.

The two branches of (6.5) and (6.6) are not applicable everywhere. The first branch of (6.5) and (6.6) applies when $q_1 + q_2 < 1{,}800$ holds, the second when $q_1 + q_2 \geq 1{,}800$ does.

In order to interpret the inequalities in terms of competitors' supplies, substitute from the first branches of (6.5) and (6.6) in $q_1 + q_2 < 1{,}800$. This results in $q_2 < 1{,}600$ and $q_1 < 1{,}600$ respectively. Likewise, substitute from the second branches of (6.5) and (6.6) in $q_1 + q_2 \geq 1{,}800$ to obtain $q_2 \geq -2{,}400$ and $q_1 \geq -2{,}400$ respectively. As competitors' outputs must be positive, these latter conditions do not constrain anything.

So, one can rephrase (6.5) and (6.6) to

$$q_1(t+1) = \begin{cases} 1{,}000 - \frac{1}{2}q_2(t) & q_2(t) < 1{,}600 \\ 3{,}000 - \frac{1}{2}q_2(t) & q_2(t) \geq -2{,}400 \end{cases} \tag{6.7}$$

$$q_2(t+1) = \begin{cases} 1{,}000 - \frac{1}{2}q_1(t) & q_1(t) < 1{,}600 \\ 3{,}000 - \frac{1}{2}q_1(t) & q_1(t) \geq -2{,}400 \end{cases} \tag{6.8}$$

Further, the supplies of the firms themselves have to be positive in order to make sense. From the first branches of (6.7) and (6.8) this means $q_2 < 2{,}000$ and $q_1 < 2{,}000$.

As the branch conditions already are constrained to $q_1, q_2 < 1{,}600 < 2{,}000$, the first branches do not become more constrained. For the second branches of (6.7) and (6.8), one finds $q_2 < 6{,}000$ and $q_1 < 6{,}000$ as conditions for positivity, which in fact does constrain more than stated in (6.7) and (6.8). Considering all the constraints, one can now write:

$$q_1(t+1) = \begin{cases} 1{,}000 - \frac{1}{2}q_2(t) & 0 < q_2(t) < 1{,}600 \\ 3{,}000 - \frac{1}{2}q_2(t) & 0 < q_2(t) < 6{,}000 \end{cases} \tag{6.9}$$

$$q_2(t+1) = \begin{cases} 1{,}000 - \frac{1}{2}q_1(t) & 0 < q_1(t) < 1{,}600 \\ 3{,}000 - \frac{1}{2}q_1(t) & 0 < q_1(t) < 6{,}000 \end{cases} \tag{6.10}$$

One issue remains to be dealt with. Both expressions for local optima in (6.9)–(6.10) have a common range of validity, $0 < q_1, q_2 < 1{,}600$. To find the global optimum one must calculate profits, i.e., just revenues, as there are no costs, according to both expressions within this range.

Consider (6.1), where $Q = q_1 + q_2$, and substitute from the first branch of (6.9) and (6.10) for firm i, to obtain its estimate of market price. This, of course depends on the supply of the other firm j, $i \neq j$,

$$p(t+1) = 100 - \frac{1}{20}\left(\left(1{,}000 - \frac{1}{2}q_j(t)\right) + q_j(t)\right) = 50 - \frac{1}{40}q_j(t) \tag{6.11}$$

Profits, using the first branch of (6.9) and (6.10) for q_i again, then become

$$\begin{aligned} p(t+1)q_i(t+1) &= \left(50 - \frac{1}{40}q_j(t)\right)\left(1{,}000 - \frac{1}{2}q_j(t)\right) \\ &= \frac{1}{80}\left(2{,}000 - q_j(t)\right)^2 \end{aligned} \tag{6.12}$$

Similarly, considering the second branch of the demand function, the expected prices would be

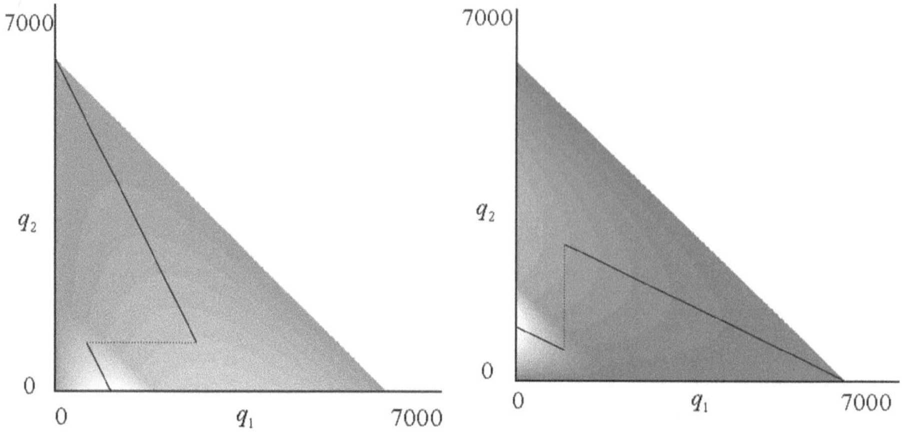

Fig. 6.2 Reaction functions for first firm (*left*) and second firm (*right*) for Palander's first case. Profit level (the brighter, the higher) indicated by *shading*

$$p(t+1)=\frac{100}{7}-\frac{1}{420}\left(\left(3{,}000-\frac{1}{2}q_j(t)\right)+q_j(t)\right)=\frac{50}{7}-\frac{1}{840}q_2(t) \tag{6.13}$$

Expected profits then, using the second branch of (6.9) and (6.10) for q_i, become

$$\begin{aligned} p(t+1)q_i(t+1)&=\left(\frac{50}{7}-\frac{1}{840}q_j(t)\right)\left(3{,}000-\frac{1}{2}q_j(t)\right) \\ &=\frac{1}{1{,}680}\left(6{,}000-q_j(t)\right)^2 \end{aligned} \tag{6.14}$$

Equating (6.12)–(6.14), the break even points are obtained,

$$q_j(t)=1{,}800\pm 200\sqrt{21}$$

Only the smaller root makes sense, as only this is in the common validity range $0<q_j(t)<1{,}600$.

So, assembling the pieces,

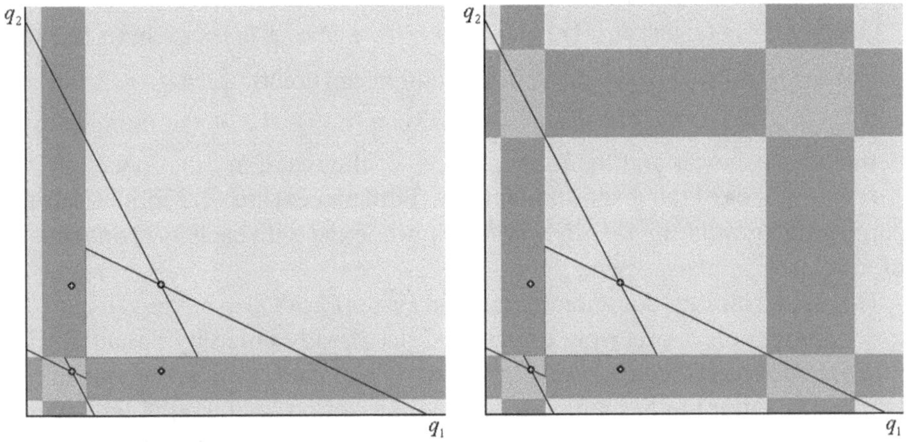

Fig. 6.3 Attractors (two fixed points, and one 2-period orbit) with basins in Palander's first model. Palander's own characterization (1936) to the *left*; as completed by modern means to the *right*

$$q_1(t+1) = \begin{cases} 1{,}000 - \frac{1}{2}q_2(t) & 0 < q_2(t) < 1{,}800 - 200\sqrt{21} \\ 3{,}000 - \frac{1}{2}q_2(t) & 1{,}800 - 200\sqrt{21} \le q_2(t) < 6{,}000 \\ 0 & 6{,}000 \le q_2(t) \end{cases} \tag{6.15}$$

$$q_2(t+1) = \begin{cases} 1{,}000 - \frac{1}{2}q_1(t) & 0 < q_1(t) < 1{,}800 - 200\sqrt{21} \\ 3{,}000 - \frac{1}{2}q_1(t) & 1{,}800 - 200\sqrt{21} \le q_1(t) < 6{,}000 \\ 0 & 6{,}000 \le q_1(t) \end{cases} \tag{6.16}$$

are the final reaction functions. These piecewise linear reaction functions are displayed in Fig. 6.2 against a background of the profit surfaces in terms of shading; brighter shade indicating higher profits.

In Fig. 6.3 the two reaction functions are displayed in the same picture quite as Palander did.

Identifying $q_1(t+1) = q_1(t) = q_1$ and $q_2(t+1) = q_2(t) = q_2$ in (6.15) and (6.16) one finds two coexistent Cournot equilibrium points, $q_1 = q_2 = 2{,}000/3$ and $q_1 = q_2 = 2{,}000$. These can also be seen in Fig. 6.3 as the intersections between the shorter and the longer pieces of the reaction functions.

As the slopes of the reaction functions (both pieces) are $-1/2$, both Cournot equilibrium points, are *locally stable*. They coexist with each its proper basin of attraction in phase space .

This local stability does not prevent the system from also having an oscillatory solution, more precisely of period 2, as already noted by Palander. His model hence has three *coexistent attractors*, two fixed points, and one oscillating. The latter too has a proper basin of attraction, included in Palander's original picture. The precision of his design, as shown in Fig. 6.3 (left), is amazing as it was done in 1936 with slide ruler and simple mechanical calculator.

However, Palander missed that the system might need initial steps for convergence on the immediate basins of the attractors, i.e., the preimages of the immediate basins. The true picture (shown to the right) is more fragmented; there are 25 basin pieces instead of 9. For a detailed analysis of this see Puu et al. (2002).

This is a proper place for a note. In discussing the dynamics of the iterated map according to (6.15) and (6.16), Palander, was keen on distinguishing between two systems assumed as different, called "*simultaneous adjustment*", and "*alternative adjustment*", and thought of this as an important distinction for the statement of dynamical systems in general. In the first both competitors react simultaneously each period according to rules such as stated in (6.15) and (6.16), in the latter they take turns in reacting every second period.

However, there is no actual difference. The dynamics is exactly *the same* in *both* of Palander's adjustment schemes. To see this, call the map stated in (6.15) $q_1(t+1) = \phi_1(q_2(t))$ and the map stated in (6.16) $q_2(t+1) = \phi_2(q_1(t))$.

Then the compositions $q_1(t+2) = \phi_1 \circ \phi_2(q_1(t))$, $q_2(t+2) = \phi_2 \circ \phi_1(q_2(t))$ are two independent well defined iterations *in each variable alone* though with a reaction period twice as long as the basic period of the model. The orbit of each such iteration from an initial condition is *uniquely* defined, one only needs initial conditions for *two* subsequent periods to produce a complete time series. The events in two dimensional phase space then only represent pairing such orbits for q_1 and q_2 together. All these issues are completely sorted out in Bischi et al. (2000).

Palander's Second Case

Demand and Cost Functions

As an alternative, Palander in his Swedish article but not in the Cowles Commission abstract, suggested:

$$p = f(Q) = \begin{cases} 100 - \dfrac{Q}{10} & Q < 900 \\ \dfrac{500}{41} - \dfrac{Q}{410} & Q \geq 900 \end{cases} \tag{6.17}$$

where the first firm has a constant marginal cost equal to $c_1 = 6$, whereas the second firm still has zero $c_2 = 0$. The firms are now asymmetrical, which has consequences.

Reaction Functions

By the same procedure as before, not repeated as it was rather lengthy but elementary, this time not forgetting to subtract costs from the revenues for the first firm, the reaction functions are

$$q_1(t+1) = \begin{cases} 470 - \dfrac{1}{2} q_2(t) & 0 < q_2(t) < 900 - 40\sqrt{41} \\ 1270 - \dfrac{1}{2} q_2(t) & 900 - 40\sqrt{41} \leq q_2(t) < 2540 \\ 0 & 2540 \leq q_2(t) \end{cases} \tag{6.18}$$

$$q_2(t+1) = \begin{cases} 500 - \dfrac{1}{2} q_1(t) & 0 < q_1(t) < 900 - 100\sqrt{41} \\ 2500 - \dfrac{1}{2} q_1(t) & 900 - 100\sqrt{41} \leq q_1(t) < 5000 \\ 0 & 5000 \leq q_1(t) \end{cases} \tag{6.19}$$

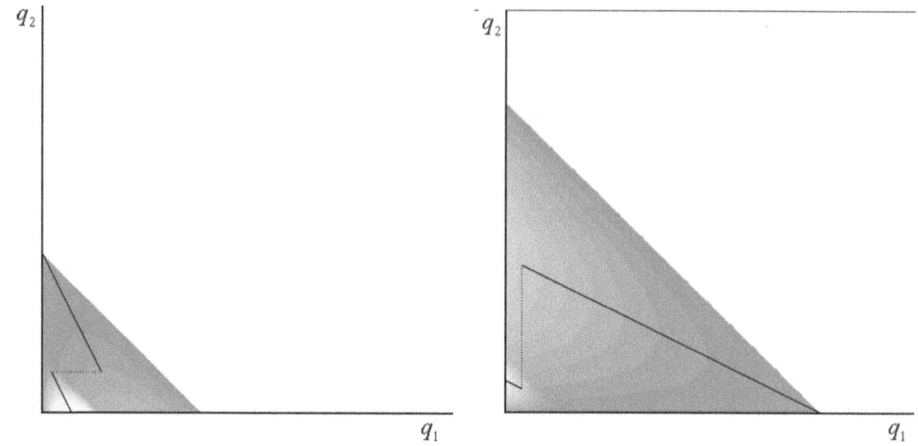

Fig. 6.4 Palander's second example. Reaction functions in duopoly of two asymmetric competitors

The reaction functions are displayed in Fig. 6.4, again against the background of the profit surfaces indicated by shading. Unlike Fig. 6.2 they are not just identical with respect to rotation around the diagonal.

It is easy to check that the reaction functions do not intersect, so *there exist no Cournot equilibria at all*. Palander noted the existence of a stable 3-period cycle, and provided a rather rudimentary picture of its attraction basin.

However, there also is a coexistent 6-cycle. Again Bischi et al. (2000) give the general facts about this case. The basin, shown in Fig. 6.5, is quite complex.

Table 6.1 Numerically obtained 3- and 6-period cycles

3-period cycle		6-period cycle	
q_1	q_2	q_1	q_2
280.6349	454.9206	280.6349	378.7302
242.5396	2,359.6825	280.6349	2,359.6825
90.1587	378.7302	90.1587	2,359.6825
280.6349	454.9206	90.1587	454.9206
242.5396	2,359.6825	242.5396	454.9206
90.1587	378.7302	242.5396	378.7302
280.6349	454.9206	280.6349	378.7302

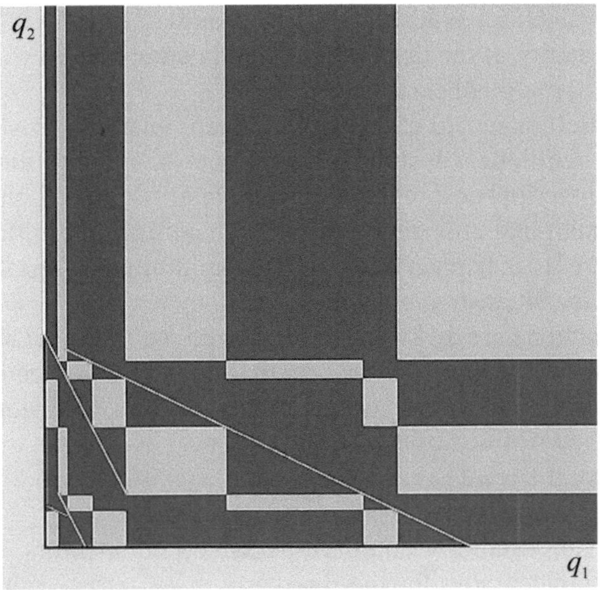

Fig. 6.5 Basins of coexistent 3-period (*dark*) and 6-period (*bright*) cycles in Palander's second duopoly model

Palander displayed one shaded area, which, in fact, is a mixture of the basins for the two coexistent periodic orbits. Probably he checked this area and noted that the system went to oscillatory motion, whereas he left the rest of the plane unchecked.

It is interesting to note that both in the 3-cycle and in the 6-cycle it is always the same three values for the q_1-coordinate and the q_2-coordinate that are visited over and over.

The way the 3-cycle becomes a 6-cycle is by each competitor lingering two periods at the same supply. It then takes six periods to arrive at the same combination of both coordinate values.

It is noteworthy that, whereas in the 3-cycle the points are not located on the reaction curves, in the 6-cycle they in fact are. So, though assuming what Palander calls "simultaneous" adjustment, the process itself settles at an "alternative" adjustment in his terminology. See Table 6.1.

Instability

In these examples by Palander, any Cournot equilibria, even in case of multiple attractors, are locally stable. This is due to the fact that the downward slope of the reaction functions is always less than unity. The reason for this, of course is the linearity of the demand function in combination with constant (zero or nonzero) marginal costs.

Palander himself sketched cases with quadratic marginal cost functions, quite as in the traditional illustrations in microeconomics textbooks. If the intersection of marginal cost with marginal revenue is located on the falling section of the marginal cost curve, then the slope of the resulting reaction function is raised. It can happen that a Cournot equilibrium is destabilised, and gives way to some interesting dynamics.

Quadratic functions are awkward to work with, so in Puu et al. (2002) it was assumed that the whole process occurred on the falling section of marginal cost, and this falling section was approximated by a downsloping line. It could then be shown how the destabilised Cournot equilibrium was replaced by orbits of very different periodicities, often coexistent, and even chaotic attractors. For the details see Sushko et al. (2002).

Another way to destabilise equilibria is adding competitors. Theocharis (1959) has been celebrated for the discovery that Cournot equilibrium in oligopoly becomes neutrally stable with three competitors, and is destabilised with four. The given format is a linear demand function and constant marginal costs. Palander's demand function is piecewise linear, but the same facts hold for this more complicated setting. With four competitors quite complex attractors can be found numerically in Palander's model.

It has been pointed out elsewhere in this book that Palander was fully aware of this destabilisation, and gave an analysis of this 20 years before Theocharis, in the very articles that have been discussed in this chapter.

Operating Several Plants

Palander also gave examples of multistability that arises on the cost side. Suppose each competitor has two units (plants, machines) they can operate, one by one, say appropriate for small scale production and large scale production respectively, or jointly, dividing the load between them according to the principle of equal marginal costs.

Then, as a rule, there arise three possibilities: Depending on the demand to be satisfied, the competitor chooses one or another, and if demand grows even more, then it will operate both. So, with two plants, there are three different production arrangements, each of which may be most advantageous in its proper range of output. See Tramontana et al. (2009).

As a consequence there may arise discontinuities in the marginal cost function quite as was the case with the marginal revenue function in the previous sections. Thus multiple local profit maxima can exist, and switching between them occurs, depending on the supply aimed at.

In the duopoly setting this planned supply depends on the expected supply by the competitor, so the switching can again give rise to discontinuous reaction functions and hence to multiple Cournot equilibria.

To avoid the complexity of three different ranges and two discontinuities, Palander assumed that the units to operate were identical. Then there remain only two options: To activate only one, i.e., any of the identical units, or both together. In the latter case marginal costs rise at half the rate at which they rise for each single unit, as output is equally divided between the units. But Palander assumed there was a discontinuous activating cost when a unit was set up for production. Palander used two different kinds of fixed costs, fixed costs proper for the equipment owned by the firm, and an activating cost for each unit of equipment. Obviously the first kind accrues no matter whether a unit is in actual use or not, so it does not influence production decisions whenever capital is fixed through some past investment decision. His second category, activation costs, seem to be most relevant, but it cannot be caught by any of the smooth production functions in common use, and have thus dropped out from the discussion. [1]

Further, Palander assumed that the two competitors were identical. Under these assumptions he described three scenarios. In two of those there are three coexistent Cournot equilibra, one symmetric with both competitors operating either one or two plants, and two asymmetric ones where one competitor operates two plants, the other one. In addition there are three two-period oscillatory solutions. In one of the described cases only the asymmetric Cournot equilibria and a two-period oscillation exist. As the effect on the scenarios of one parameter, activation cost, is studied, a sequence of bifurcations is implied. Given this Palander's discussion has a very modern aspect.

[1] The present author once visited a window glass factory in France where a huge machine poured out a continuous sheet of glass. It was designated for being cut in pieces and put on a number of polishing machines. However, one of the polishing machines was in repair, so the present author witnessed the seemingly absurd scenery of workmen smashing the glass coming out from the the far end of the big machine and driving it back to be inserted anew. It was simply too expensive to shut off the big machine and start it up again while the small was repaired. This illustrates Palander's distinction, very relevant in practice, but impossible to catch by any smooth production function.

The Cost Function

The assumed cost functions for each unit are quadratic, hence the marginal cost functions are linear,

$$C_i(q_i) = a + 10q_i + \tfrac{1}{10}q_i^2 \quad i = 1,2 \tag{6.20}$$

where a denotes the activation cost. Marginal costs are hence

$$MC_i(q_i) = 10 + \tfrac{1}{5}q_i \quad i = 1,2 \tag{6.21}$$

When the competitors operate two plants, then, given the plants are identical, they produce $\frac{1}{2}q_i$ in each. Substituting $\frac{1}{2}q_i$ for q_i in (6.20), and multiplying by 2, the cost function reads

$$C_i(q_i) = 2a + 10q_i + \tfrac{1}{20}q_i^2 \quad i = 1,2 \tag{6.22}$$

Note that the linear term is not changed, whereas the quadratic term is halved. Further the activation cost is doubled as both units are activated. To (6.22) belongs a marginal cost function,

$$MC_i(q_i) = 10 + \tfrac{1}{10}q_i \quad i = 1,2 \tag{6.23}$$

Obviously, each competitor chooses the minimum of (6.20) and (6.22) for each output q_i. To find the point at which these functions break even, put

$$a + 10q_i + \tfrac{1}{10}q_i^2 = 2a + 10q_i + \tfrac{1}{20}q_i^2 \quad i = 1,2 \tag{6.24}$$

and solve for

$$q_i = \sqrt{20a} \tag{6.25}$$

Thus the combined cost function becomes

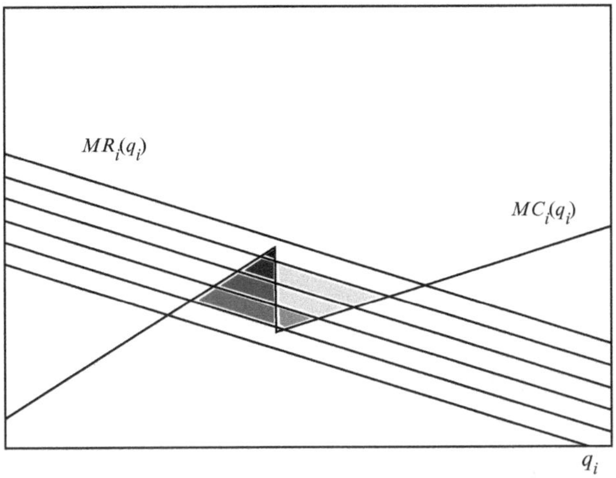

Fig. 6.6 Marginal cost function with a family of marginal revenue functions

$$C_i(q_i) = \begin{cases} a + 10q_1 + \frac{1}{10}q_i^2 & q_i \le \sqrt{20a} \\ 2a + 10q_1 + \frac{1}{20}q_i^2 & q_i > \sqrt{20a} \end{cases} \quad i = 1,2 \tag{6.26}$$

This cost function is continuous, and piecewise smooth, but the corresponding marginal cost function

$$MC_i(q_i) = \begin{cases} 10 + \frac{1}{5}q_1 & q_i \le \sqrt{20a} \\ 10 + \frac{1}{10}q_1 & q_i > \sqrt{20a} \end{cases} \quad i = 1,2 \tag{6.27}$$

obviously becomes discontinuous in the point $q_i = \sqrt{20a}$, as illustrated in Fig. 6.6. Note that the marginal cost does not depend on the parameter a.

The Reaction Functions

As in the first line of (6.6), the inverse demand function is assumed to be

$$p = 100 - \frac{1}{20}(q_1 + q_2) \tag{6.28}$$

The corresponding marginal revenue functions then are

$$MR_i\left(q_i,q_j\right)=100-\tfrac{1}{10}q_i-\tfrac{1}{20}q_j \quad i,j=1,2 \quad i\neq j \tag{6.29}$$

Equating (6.29) to (6.21), and solving for q_i yields

$$q_i=300-\tfrac{1}{6}q_j \quad i,j=1,2 \quad i\neq j, \tag{6.30}$$

whereas equating (6.29) to (6.23) and solving yields

$$q_i=450-\tfrac{1}{4}q_j \quad i,j=1,2 \quad i\neq j. \tag{6.31}$$

These provide local profit maxima. To choose the global maximum, one has to calculate profits,

$$\Pi_i\left(q_i,q_j\right)=100q_i-\tfrac{1}{20}\left(q_i+q_j\right)q_i-C_i\left(q_i\right) \quad i,j=1,2 \quad i\neq j. \tag{6.32}$$

Using the cost function (6.20) and substituting from (6.30) one has

$$\Pi_i\left(q_j\right)=\tfrac{1}{240}\left(1{,}800-q_j\right)^2-a \quad i,j=1,2 \quad i\neq j, \tag{6.33}$$

whereas using (6.22) and substituting from (6.31) yields

$$\Pi_i\left(q_j\right)=\tfrac{1}{160}\left(1{,}800-q_j\right)^2-2a \quad i,j=1,2 \quad i\neq j. \tag{6.34}$$

To find the discontinuity point for the reaction functions, (6.33) is equated to (6.34), and the resulting equation is solved for q_j, i.e.,

$$q_j=1{,}800-\sqrt{480a} \quad j=1,2 \tag{6.35}$$

The resulting reaction functions then are

$$q_i = \begin{cases} 450 - \frac{1}{4} q_j & q_j \leq 1{,}800 - \sqrt{480a} \\ 300 - \frac{1}{6} q_j & q_j > 1{,}800 - \sqrt{480a} \end{cases} \quad i, j = 1{,}2 \quad i \neq j \tag{6.36}$$

The Cournot Equilibria

The reaction functions (6.36) are illustrated in Figs. 6.7–6.9. As one can see, there are four possible candidates for Cournot equilibria, two on the diagonal, and two off the diagonal. In the first case either both firms use one production unit each, or both use two units. In the second case one firm uses one unit, another two units.

Like the reaction functions themselves, the Cournot equilibria do not depend on the parameter *a*.

They can even be easily calculated. Putting

$$q_i = q_j = q \tag{6.37}$$

in (6.30),

$$q = 300 - \frac{1}{6} q \tag{6.38}$$

is obtained and has the solution

$$q_i = q_j = \frac{1{,}800}{7} \tag{6.39}$$

Likewise, from (4.31)

$$q = 450 - \frac{1}{4} q \tag{6.40}$$

the solution

$$q_i = q_j = 360 \tag{6.41}$$

results.

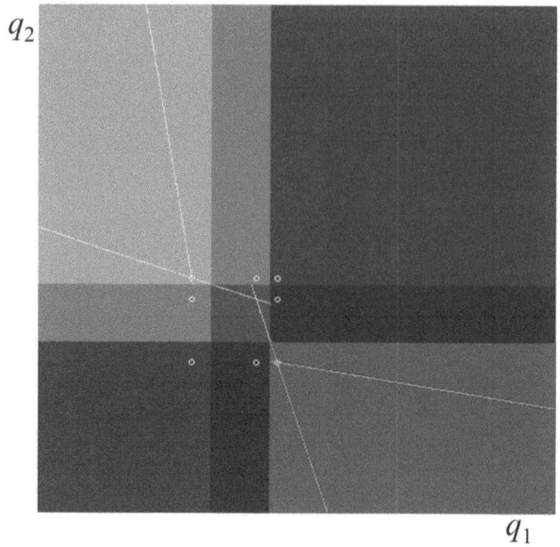

Fig. 6.7 Coexistent attractors and basins, three Cournot equilibria and three 2-period orbits. Parameter a = 4,230

Combining (6.30) and (6.31) to the system

$$\begin{cases} q_i = 300 - \frac{1}{6} q_j \\ q_j = 450 - \frac{1}{4} q_i \end{cases} \quad i, j = 1,2 \quad i \neq j \tag{6.42}$$

the solutions

$$\begin{cases} q_i = \dfrac{5,400}{23} \\ q_j = \dfrac{9,000}{23} \end{cases} \quad i, j = 1,2 \quad i \neq j \tag{6.43}$$

are obtained.

However, even if the locations of the Cournot points do not depend on the parameter a, the parameter decides whether they lie on the relevant sections of the discontinuous reaction functions or not. Palander described three scenarios, with $a = 4,230$ displayed in Fig. 6.7, $a = 4,500$ in Fig. 6.8, and $a = 5,000$ in Fig. 6.9.

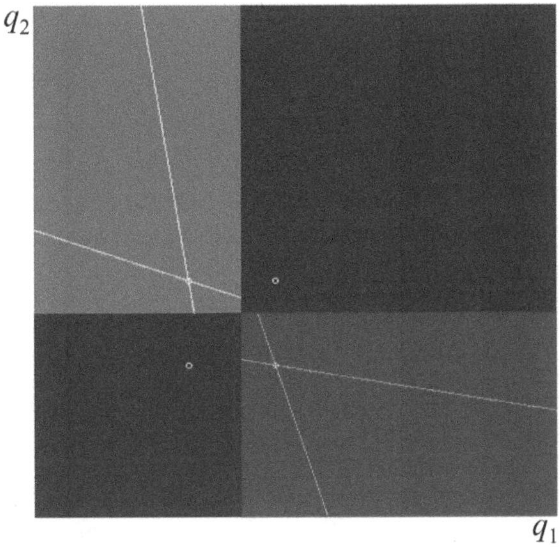

Fig. 6.8 Coexistent attractors and basins, two Cournot equilibria and one 2-period orbit. Parameter a = 4,500

In the first case the higher symmetric (6.41) equilibrium coexists with the two asymmetric (6.43) equilibria; in the last case the lower symmetric (6.39) equilibrium coexists with the asymmetric (6.43), and in the intermediate case only the two asymmetric equilibria exist.

Attractors and Basins

In Figs. 6.7–6.9, Palander's three cases are illustrated. The pictures contain the reaction functions. As mentioned, their shapes are independent of the bifurcation parameter *a*, i.e., the activation cost for the production unit. However, the switching point (6.35), and therefore the segments of them representing global profit maxima do depend on the parameter. Hence, the existence of the four possible Cournot equilibria also depends on the parameter.

Due to the different parameter values, Fig. 6.7 shows three coexistent Cournot equilibrium points, one symmetric and two asymmetric. In Fig. 6.8 just the two asymmetric ones exist, whereas in Fig. 6.9, there is again a symmetric equilibrium, but now the lower one where both competitors use only one production unit. In addition one can imagine cases, nor studied by Palander, where only one symmetric Cournot equilibrium, the lower or the higher, but no asymmetric ones exist.

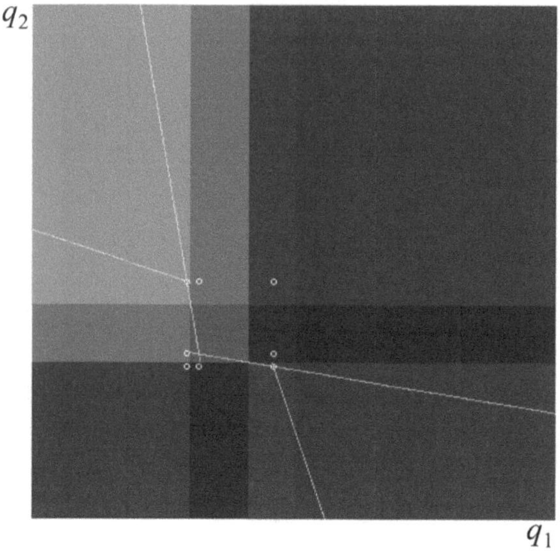

Fig. 6.9 Coexistent attractors and basins, three Cournot equilibria and three 2-period orbits. Parameter a = 5,000

It is worth noting that, due to the low slopes of the reaction functions, the Cournot equilibria are always stable when they exist. From (6.36) the highest slope is $-\frac{1}{4}$, and therefore the absolute value of the Jacobian is bounded by $\frac{1}{16}$ from above. This is lower than in the case of constant marginal costs, which depends on Palander's assumption of a positive quadratic term in the cost functions (6.20).

Stability means that the fixed points, whenever they exist, always have a basin of attraction. Again, Palander was amazingly accurate in characterising these basins.

In addition to the fixed points, the model also produces oscillatory solutions. Palander noted this, but again distinguishing between simultaneous adjustment and sequential adjustment, he thought some of these would eventually land in any of the fixed points depending on which competitor starts the process.

This is not quite correct. In Figs. 6.7 and 6.9 one can see three additional 2-period oscillatory attractors, one located on the diagonal, where the competitors move in phase, and two asymmetric ones where the competitors move out of phase. In Fig. 6.8 there is just one 2-period oscillatory solution along the diagonal, the duopolists moving in phase.

It is obvious that all points of the attractors are located on a square grid of three or two intersecting horizontal and vertical lines, quite as analysed in Bischi et al. (2000) for systems of this type.

Due to the stability of the existent Cournot equilibria, all bifurcations occur due to disappearing or emerging basins. Palander's sequence displays two of them. The exact bifurcation values for the parameters can be obtained considering that the break point coordinate (6.35) coincides with one of the symmetric Cournot equilibrium points (6.39) or (6.41).

Equating (6.35)–(6.39),

$$\frac{1{,}800}{7} = 1{,}800 - \sqrt{480a} \quad j = 1{,}2\,, \tag{6.44}$$

whereas equating (6.35)–(6.41),

$$360 = 1{,}800 - \sqrt{480a} \quad j = 1{,}2 \tag{6.45}$$

is obtained. These equations can then be solved for the exact bifurcation values,

$$a = \frac{243{,}000}{49} \approx 4959\,, \tag{6.46}$$

$$a = 4{,}320\cdot \tag{6.47}$$

There remain interesting issues to study, for instance, relaxing the restrictive assumption that the production units are identical. This implies, as already mentioned, that the firms have three sections of the marginal cost function. This issue was dealt with in Tramontana et al. (2009), though in a different setting; with isoelastic demand and CES generated cost functions with fixed capital. Unfortunately, the symmetric case studied does not permit the coexistence of Cournot equilibria, which Palander's original model does.

References

Arrow, K. J., & Debreu, G. (1954). Existence of equilibrium for a competitive economy. *Econometrica, 22*, 265–290.

Bischi, G. I., Mammana, C., & Gardini, L. (2000). Multistability and cyclic attractors in duopoly games. *Chaos, Solitons & Fractals, 11*, 543–564.
Palander, T. (1936). Instability in competition between two sellers, in Abstracts of Papers Presented at the Research Conference on Economics and Statistics held by the Cowles Commission at Colorado College, *Colorado College Publications*, General Series No. 208, Studies Series No. 21.
Palander, T. (1939). Konkurrens och marknadsjämvikt vid duopol och oligopol. *Ekonomisk Tidskrift, 41*, 124–145, 222–250.
Puu, T., Gardini, L., & Sushko, I. (2002). Cournot duopoly with kinked demand according to Palander and Wald. In T. Puu, & I. Sushko, (Eds.), *Oligopoly dynamics* (pp. 111–146). Berlin: Springer.
Robinson, J. (1933). *The Economics of Imperfect Competition,* London: MacMillan.
Sushko, I., Gardini, L., & Puu, T. (2002). Duopoly with piecewise linear discontinuous reaction functions. In T. Puu, & I. Sushko, (Eds.), *Oligopoly dynamics*, (pp. 147–169). Berlin: Springer.
Sweezy, P. M. (1939). Demand under conditions of oligopoly. *Journal of Political Economy, 47*, 568–573.
Theocharis, R. D. (1959). On the stability of the Cournot solution of the oligopoly problem. *Review of Economic Studies, 27*, 133–134.
Tramontana, F., Gardini, L., & Puu, T. (2009). Cournot duopoly when the competitors operate several plants. *Journal of Economic Dynamics and Control, 33*, 250–265.
Wald, A. (1936). Über einige Gleichungssysteme der mathematischen Ökonomie. *Zeitschrift für Nationalökonomie,* Dezember *7*, 637–670.

7 The Hotelling Paradox

Spatial Oligopoly

As mentioned in Chap. 1, the so called Bertand variant of oligopoly, with differentiated goods, is a less yielding format for interesting models than the very clear cut original Cournot case. The main reason is the difficulty of quantifying the difference between close substitutes. There is, however, one exception, and that is the 1929 model by Harold Hotelling. In this frame consumers regard the commodity supplied by different competitors as homogenous, but there is a difference between suppliers due to spatial distance from the consumers and the necessary transportation costs. Each producer has its market area, within which it is a monopolist. As the consumers always buy from the cheapest supplier, in terms of mill price plus transportation cost, there is competition at the market boundaries defined through the condition that supply prices break even. If one local monopolist raises mill price, it will see the market area diminished, and with this total demand for its supply.

This general idea is most fruitful, though its potential has been drowned in a paradox which also was suggested in the short Hotelling paper: If space is a finite one-dimensional line segment, and there are two competitors located in this segment, then, provided they can both choose location as well as mill price, the optima will be indeterminate. They will thus both locate in the middle of the segment, in stead of at the midpoints of each half-segment which would be socially optimal to the purpose of minimizing total transportation cost.

This paradox hides two different problems. One is the possibility of cutting out the competitor and taking the entire market, which remains a possibility for global optimum under any assumptions. But this is just the same problem as pointed out by Cournot's critics; with both competitors in the same location nothing of the particularly spatial issues remains.

T. Puu, *Oligopoly: Old Ends – New Means*, DOI 10.1007/978-3-642-15964-0_7,

A different problem is that under Hotelling's specific assumptions there do not exist any local inner optima, because the optimization problem is linear in the choice of location. In the lack of any inner optima with central location for each competitor in its proper market area, one is naturally thrown in the arms of the paradoxical cutting out situation as the only possibility.

So, what are Hotelling's assumptions? As a matter of fact they are slightly contradictory. The boundaries of the market areas are established on the basis that no consumers buy from a supplier whose price including transportation cost is higher. Consumers are hence sensitive to price. But, on the other hand they always buy a fixed quantity independent of price. This leads to the linearity of the optimand with respect to location and the nonexistence of inner optima.

In his verbal discussion Hotelling indicated that making demand elastic to price would remove the problem, but he did not formalize this. Somewhat later, Lerner and Singer (1937) in an ingenious graphical analysis actually showed that just the assumption of a reservation price, above which no consumer would buy the commodity, removed the indeterminacy.

And Smithies (1941a, b) proposed a formal model using a linear demand function which he claimed would make the local optima determinate. But at the same time he argued that the integrations involved would be so messy that analysis would be almost impossible. Smithies's formulation agrees with what Hotelling actually had in mind, but, as will be shown below, his fear for the technical difficulty of the analysis was much exaggerated.

It can, in fact, be shown formally that inner location and pricing optima exist, and one can even devise a dynamical stepwise search system for location and pricing that is convergent to location of the firms at the centres of each its monopoly area.

However, the cutting out possibility remains as a paradoxical possibility in the global sense, even when the inner solutions exist. One should perhaps also add that it would be interesting to know these facts in true two-dimensional geographical space. However, as always, spatial analysis in economics stops after the simplification to one dimension. In two dimensions, Smithies definitely is right, the integrations over market areas bounded by more or less complex boundaries become just too messy. Further, the present author has seen many attempts at generalizing the Hotelling problem to two dimensions, unfortunately none of them convincing. It is even difficult to imagine what the Hotelling problem generalized to two dimensions is. Perhaps three competitors in an equilateral triangle, or something similar. Anyhow, too complicated for formal analysis.

Hotelling assumed a bounded line segment on which two competitors located. However, to approach this problem it is better to first consider the issue of a train of firms located on a line with no fixed endpoints where each competitor has a neighbour to the right and to the left.

A Firm Between Competitors

Consider a firm located along a one dimensional line at point x_i. To the left of it there is another firm at point x_{i-1}, to the right of it another at point x_{i+1}.

The firms charge mill prices p_{i-1}, p_i, p_{i+1}. Mill pricing means that the consumers pay for transportation; the good becomes more expensive farther away from the producing firm. Other price policies, such as discriminatory pricing, including uniform delivered price, are conceivable as each firm is a local monopoly and competition is confined to the boundary points.

Local demand at some given point x for the good transported from the firm at x_i is:

$$f(p) = \begin{cases} a - bp & p < \frac{a}{b} \\ 0 & p \geq \frac{a}{b} \end{cases} \tag{7.1}$$

where

$$p = p_i + k|x - x_i| \tag{7.2}$$

Note that, unlike the convention in previous chapters, the demand function is not written in inverse form. Hence, it is *a/b*, not *a*, that denotes maximum chargeable local price *p*, mill price accrued by transportation costs. In Hotelling's original setting $b = 0$, so demand is constant and independent of price.

Further, note that (7.2) means transportation costs that increase linearly with distance, thus deviating from the rather popular but construed d'Aspremont et al. (1979) assumption of a quadratic dependence, invented to cope with the assumption $b = 0$. Transportation cost rate per unit of distance is denoted k.

Total demand for the *ith* firm is

$$Q_i = \int_{l_i}^{r_i} f\left(p_i + k\left|x - x_i\right|\right)dx \tag{7.3}$$

The symbols l_i, r_i denote the left and right market boundary points for the *ith* firm. For the linear demand function (7.1) one can evaluate the closed form integral (7.3),

$$Q_i = \left(a - bp_i\right)\left(r_i - l_i\right) - \frac{bk}{2}\left(\left(x_i - l_i\right)^2 + \left(r_i - x_i\right)^2\right) \tag{7.4}$$

The market boundary points are not fixed, but determined by the conditions that local supply prices from different suppliers be equal, i.e.,

$$p_i + k\left(x_i - l_i\right) = p_{i-1} + k\left(l_i - x_{i-1}\right) \tag{7.5}$$

$$p_i + k\left(r_i - x_i\right) = p_{i+1} + k\left(x_{i+1} - r_i\right) \tag{7.6}$$

The conditions (7.5) and (7.6) can be solved for:

$$l_i = \frac{x_{i-1} + x_i}{2} + \frac{p_i - p_{i-1}}{2k} \tag{7.7}$$

$$r_i = \frac{x_i + x_{i+1}}{2} + \frac{p_{i+1} - p_i}{2k} \tag{7.8}$$

The notation conventions can be seen in Fig. 7.1. These imply that always $l_i = r_{i-1}$ and $r_i = l_{i+1}$. Substituting from (7.7) and (7.8) in the integral (7.4),

$$\begin{aligned} Q_i &= \left(a - bp_i\right)\frac{1}{2k}\left(k\left(x_{i+1} - x_{i-1}\right) + p_{i+1} + p_{i-1} - 2p_i\right) \\ &- \frac{b}{8k}\left(\left(kx_{i-1} - kx_i + p_i - p_{i-1}\right)^2 + \left(kx_{i+1} - kx_i + p_{i+1} - p_i\right)^2\right) \end{aligned} \tag{7.9}$$

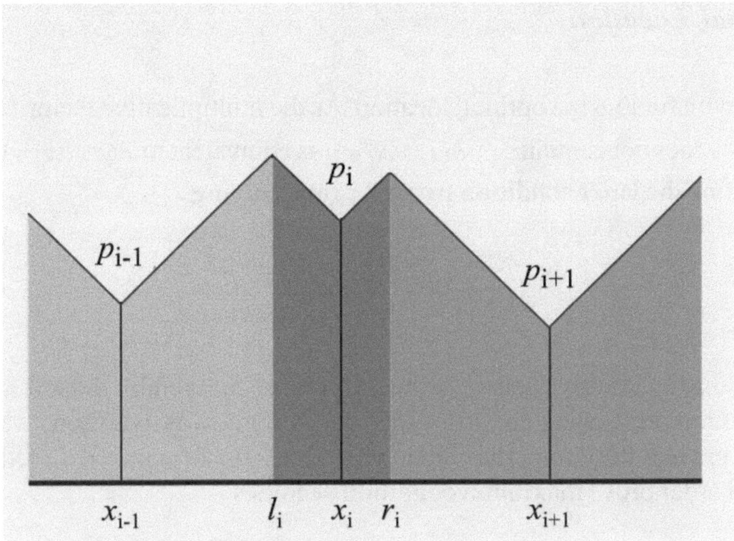

Fig. 7.1 Location and mill price for an intermediate firm

The *i*th firm will maximize its profits, i.e., its revenue which equals mill price p_i multiplied by total demand in its market area according to (7.9) minus total cost which again can be taken as constant per unit output, $C_i(Q_i) = c_i Q_i$. Hence,

$$\Pi_i = (p_i - c_i) Q_i \tag{7.10}$$

is the maximand, where (7.9) is to be substituted. The firm can choose its mill price p_i *and* its location x_i. Note that profits (7.10), with (7.9) substituted, is a quadratic function of x_i and a cubic function of p_i. In Hotelling's original setting, with $b = 0$, profits become quadratic in mill price and linear in location. Maximizing linear functions never leads to any nice inner solutions, so, one can understand the reason for inventing higher order transportation costs than motivated by intuition, even if it seems better to just stick to elastic demand. Further, note that the integrations were quite straightforward and by no means as messy as claimed by Smithies (1941a, b).

Optimal Location

Begin with finding the optimal location. As the multiplicative factor $(p_i - c_i)$ in (7.10) does not contain x_i, $\partial \Pi_i / \partial x_i = 0$ is equivalent to $\partial Q_i / \partial x_i = 0$. Thus, evaluating the latter condition from (7.9) and solving,

$$x_i = \frac{x_{i-1} + x_{i+1}}{2} + \frac{p_{i+1} - p_{i-1}}{2k} \tag{7.11}$$

This makes intuitive sense: If the mill prices charged by the left and right competitors are equal, then the firm locates midway between, whereas a higher price of the competitor left or right drags the firm in that direction. The second order profit maximum condition becomes

$$\frac{\partial^2 \Pi_i}{\partial x_i^2} = -\tfrac{1}{2} bk(p_i - c_i) < 0 \tag{7.12}$$

Recall that in the original setup of the model by Hotelling (1929), where $b = 0$ (and $c_i = 0$) was assumed, there were no inner solutions.

Using (7.11), the expressions (7.7) and (7-8) now become

$$l_i = \frac{3x_{i-1} + x_{t+1}}{4} - \frac{3p_{i-1} - 2p_i - p_{i-1}}{4k} \tag{7.13}$$

$$r_i = \frac{x_{i-1} + 3x_{t+1}}{4} - \frac{p_{i+1} + 2p_i - 3p_{i-1}}{4k} \tag{7.14}$$

From (7.11), (7.13) and (7.14) one gets the expression

$$(x_i - l_i) = (r_i - x_i) = \frac{x_{i+1} - x_{i-1}}{4} + \frac{p_{i-1} - 2p_i + p_{i+1}}{4k} = \tfrac{1}{2} d_i \tag{7.15}$$

which may be defined as the market radius, i.e., half the market diameter d_i.

Hence, even if the firm does *not* locate halfway between its neighbours, it *does* locate *halfway between the boundary points*, which means in the centre of its proper market interval $[l_i, r_i]$. As a consequence consumer prices are equal at both ends of the market.

It is now possible to simplify the equation for total demand (7.9) through substitutions from (7.11), (7.13), and (7.14). To find a succinct expression for (7.9), define the auxiliary variable

$$\lambda_i = p_{i-1} + p_{i+1} + k\left(x_{i+1} - x_{i-1}\right) \tag{7.16}$$

Note that λ_i only depends on p_{i-1} and p_{i+1}, but *not* on p_i. Hence, in further optimization with respect to mill price the auxiliary variable λ_i can be treated as a constant.

Using (7.16) one gets

$$Q_i = \frac{b}{16k}(\lambda_i - 2p_i)\left(8\left(\frac{a}{b} - p_i\right) - (\lambda_i - 2p_i)\right) \tag{7.17}$$

Optimal Pricing

Substitute from (7.17) in (7.10) to obtain

$$\Pi_i = \frac{b}{16k}(p_i - c_i)(\lambda_i - 2p_i)\left(8\frac{a}{b} - \lambda_i - 6p_i\right) \tag{7.18}$$

Profit according to (7.18) is a cubic in p_i. Differentiating and equating to zero hence produces a quadratic equation whose roots are

$$\begin{aligned} p_i = {} & \frac{4}{9}\frac{a}{b} + \frac{1}{3}c_i + \frac{1}{9}\lambda_i \\ & \pm \frac{1}{18}\sqrt{36\left(\frac{a}{b} - c_i\right)^2 - 12\left(\frac{a}{b} - c_i\right)\left(2\frac{a}{b} - \lambda_i\right) + 13\left(2\frac{a}{b} - \lambda_i\right)^2} \end{aligned} \tag{7.19}$$

The second derivatives of the profit function, substituting back for price from (7.19) become

$$\frac{\partial^2 \Pi_i}{\partial p_i^2} = \pm \frac{1}{4k}\sqrt{36\left(\frac{a}{b}-c_i\right)^2 - 12\left(\frac{a}{b}-c_i\right)\left(2\frac{a}{b}-\lambda_i\right) + 13\left(2\frac{a}{b}-\lambda_i\right)^2} \quad (7.20)$$

As the expressions under the root sign are the same in (7.19) and (7.20), it follows that, whenever the roots are real, the smaller one is a maximum.

The setup of the problem, given feasible parameter values for the demand and cost structure, hence guarantees the existence of well defined inner optima for both mill price and location.

The bits and pieces now at hand are sufficient for defining equilibrium configurations for systems of oligopolistic firms, as well as dynamic interactive processes for these. The outcomes for both problems, of course, depend on the setup of the global system one has in mind, in particular on whether the competitors crowd on some line segment with fixed endpoints as Hotelling assumed, or whether the space is unbounded; not necessarily of infinite length but for instance the circumference of a circle.

Before proceeding to different cases of this global setup, it is necessary to consider a few constraining issues.

As always, a linear demand function has non-smoothness problems where it cuts the axes. Of course, prices cannot be negative; they must even be higher than the unit production costs, i.e., $p_i > c_i$. Further, above the maximum demand price *a/b*, demand drops to zero. This condition holds for *mill price plus transportation cost.* As the latter rise with the distance from the location of the firm, the highest prices for the consumers incur at the boundary points. Hence, it is sufficient to stipulate that

$$p_i + k(x_i - l_i) \leq \tfrac{a}{b} \quad (7.21)$$

$$p_i + k(r_i - x_i) \leq \tfrac{a}{b} \quad (7.22)$$

Substituting from (7.13) and (7.14) for l_i, r_i and from (7.11) for x_i, (7.21) and (7.22) boil down to one and the same condition,

$$(p_{i-1} + 2p_i + p_{i+1}) - k(x_{i+1} - x_{i-1}) \leq 4\tfrac{a}{b} \quad (7.23)$$

Disjoint Monopolies

Should condition (7.23) not be satisfied, then the market interval for the *i*th firm becomes less than the interval available, i.e., the firm becomes a monopoly totally cut off from competition with other firms. Demand drops to zero at a distance $\frac{1}{2}d_i$ on either side of the location of the firm, i.e.,

$$a - b\left(p_i + \tfrac{1}{2}kd_i\right) = 0 \tag{7.24}$$

This equation can be solved for the market diameter d_i

$$d_i = \frac{2}{k}\left(\frac{a}{b} - p_i\right) \tag{7.25}$$

The quantity integral (7.3) and (7.4) hence becomes

$$Q_i = \int_{x_i - d_i/2}^{x_i + d_i/2} f\left(p_i + k\left|x - x_i\right|\right)dx = \left(a - bp_i\right)d_i - \tfrac{1}{4}bkd_i^2 \tag{7.26}$$

or, with substitution for the market diameter from (7.25),

$$Q_i = \frac{b}{k}\left(\frac{a}{b} - p_i\right)^2 \tag{7.27}$$

Profits then are:

$$\Pi_i = \frac{b}{k}\left(p_i - c_i\right)\left(\frac{a}{b} - p_i\right)^2 \tag{7.28}$$

Differentiating this cubic with respect to p_i, equating to zero, and solving, the two roots of the quadratic become

$$p_i = \frac{a}{b} \tag{7.29}$$

$$p_i = \frac{1}{3}\frac{a}{b} + \frac{2}{3}c_i \tag{7.30}$$

The solution (7.29) makes profits zero as one can see from (7.28), whereas (7.30) is a well known formula for monopoly pricing with linear demand. See Beckmann (1968).

Second derivatives of (7.28) for the two roots (7.29) and (7.30) become

$$\frac{\partial^2 \Pi_i}{\partial p_i^2} = \pm \frac{2b}{k}\left(\frac{a}{b} - c_i\right) \tag{7.31}$$

so, given that marginal cost must be less than maximum price, i.e., $\frac{a}{b} > c_i$, the second solution (7.30) is a profit maximum.

It is easy to calculate maximum monopoly profits through substituting from (7.30) into (7.28)

$$\Pi_i = \frac{4b}{27k}\left(\frac{a}{b} - c_i\right)^3 \tag{7.32}$$

The disjoint monopoly solution is relevant when demand drops to zero at a distance from the mill before consumer prices with accumulated transportation costs break even for adjacent competitors. The market areas of neighbouring firms no longer touch, but are isolated, possibly with intervals in between which are not served by any firm – the price would simply be too high for anybody to buy the commodity. Whether this occurs seems to be a question of how many firms crowd on a given distance, what the maximum demand price is, and what the marginal costs are.

First when the firms are squeezed together does oligopoly arise. Total space D occupied by n touching monopolies is

$$D = \sum_{i=1}^{n} d_i = \frac{2}{k}\sum_{i=1}^{n}\left(\frac{a}{b} - c_i\right) \tag{7.33}$$

Oligopoly and competition arises when the total space available D is less than as stated in (7.33)

Two Competitors on a Fixed Interval

Suppose there are just two competitors, $n = 2$, on a fixed interval, as assumed by Hotelling. For the sake of symmetry put $l_1 = -1$ and $r_2 = 1$. See Fig. 7.2. From (7.7) and (7.8) then

$$l_2 = r_1 = \frac{x_1 + x_2}{2} + \frac{p_2 - p_1}{2k} \tag{7.34}$$

From (7.4) further

$$Q_1 = (a - bp_1)(1 + r_1) - \frac{kb}{2}\left((1 + x_1)^2 + (r_1 - x_1)^2\right) \tag{7.35}$$

$$Q_2 = (a - bp_2)(1 - l_2) - \frac{kb}{2}(l_2 - x_2)^2 + \left((1 - x_2)^2\right) \tag{7.36}$$

Location

As location does not enter the multiplicative factor for profits in (7.10), the optimal locations can again be found through putting the derivatives of sales (7.35) and (7.36) with respect to locations x_1, x_2 equal to zero, and solving.

This is precisely Hotelling's problem, though, due to the elasticity of demand, inner solutions are now obtained

$$x_1 = \frac{x_2 - 4}{5} + \frac{p_2 - 3p_1}{5k} + \frac{2a}{5kb} \tag{7.37}$$

$$x_2 = \frac{x_1 + 4}{5} + \frac{3p_2 - p_1}{5k} - \frac{2a}{5kb} \tag{7.38}$$

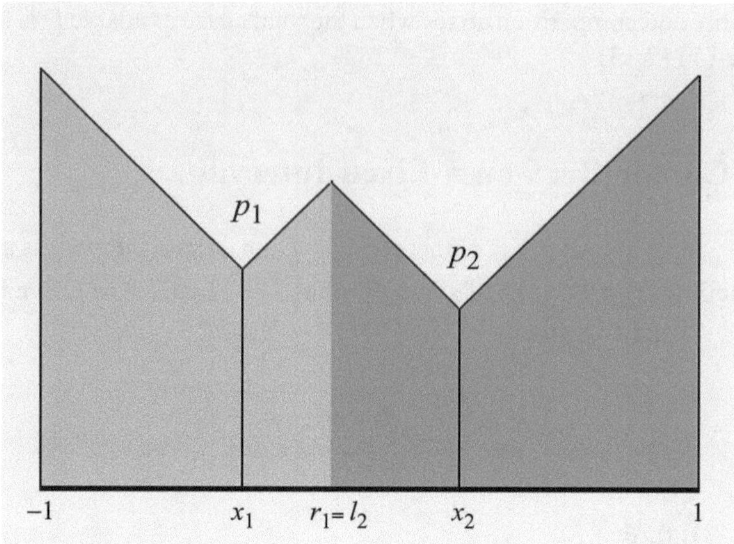

Fig. 7.2 Locations and mill prices for two firms on a fixed interval

Note that (7.37) and (7.38) are different from (7.11), as now one boundary point is fixed for each firm. Further, these expressions are relevant for best replies for the next steps in a dynamic process.

If one considers (7.37) and (7.38) as a *simultaneous system*, then one can solve explicitly for the locations in the Cournot equilibrium state,

$$x_1 = -\frac{2}{3} + \frac{a}{3kb} - \frac{2p_1 - p_2}{3k} \tag{7.39}$$

$$x_2 = \frac{2}{3} - \frac{a}{3kb} + \frac{2p_2 - p_1}{3k} \tag{7.40}$$

Adding (7.39) and (7.40) yields

$$x_1 + x_2 = \frac{p_2 - p_1}{k} \tag{7.41}$$

i.e., in case of equal mill prices, the competitors locate symmetrically around the midpoint of the fixed interval, otherwise the centre is dragged in the direction of the higher mill price.

Substituting (7.41) in (7.34)

$$l_2 = r_1 = x_1 + x_2 = \frac{p_2 - p_1}{k} \tag{7.42}$$

Subtracting (7.37) from (7.38), one obtains the distance between the equilibrium locations of the firms

$$x_2 - x_1 = \frac{4}{3} - \frac{2a}{3kb} + \frac{p_1 + p_2}{3k} \tag{7.43}$$

Hence, both competitors locate in the same point, if and only if

$$\frac{p_1 + p_2}{k} = 4 - \frac{2a}{kb} \tag{7.44}$$

holds. Clustering is thus a very special outcome once demand is elastic.

In disequilibrium, which is of interest when one wants to formulate a dynamic system, rather than to find out the facts of Cournot equilibrium, the inner market boundary point, resulting from the location choices, is *not* as stated in (7.42). The result due to the location choice of one firm or another is obtained through substituting (7.37) and (7.38) in (7.34), to obtain:

$$r_1 = l_2 = \frac{3x_2 - 2}{5} + \frac{3p_2 - 4p_1}{5k} + \frac{a}{5kb} \tag{7.45}$$

$$r_1 = l_2 = \frac{3x_1 + 2}{5} + \frac{4p_2 - 3p_1}{5k} - \frac{a}{5kb} \tag{7.46}$$

As (7.37) and (7.38) are the general expressions for locations, not necessarily in equilibrium, (7.45) and (7.46) are the same for the inner market boundary.

Pricing

Expressions (7.37), (7.38), (7.45) and (7.46) can now be substituted in (7.35) and (7.36), which can be stated by one identical expression

$$Q_1 = \frac{b}{10k}\left(6\left(\frac{a}{b}-p_1\right)^2 - 4\left(\frac{a}{b}-p_1\right)\left(\frac{a}{b}-\lambda_1\right)-\left(\frac{a}{b}-\lambda_1\right)^2\right) \tag{7.47}$$

where

$$\lambda_1 = p_2 + k(1+x_2) \tag{7.48}$$

$$\lambda_2 = p_1 + k(1-x_1) \tag{7.49}$$

Note that λ_i now have the following nice interpretation: They are the mill prices accumulated by transportation costs for the commodity transported from the firm to the right all the way to the left endpoint of the interval, and from the left firm all the way to the right endpoint.

From (7.9) one gets profits:

$$\Pi_i = (p_i - c_i)\frac{b}{10k}\left(6\left(\frac{a}{b}-p_i\right)^2 - 4\left(\frac{a}{b}-p_i\right)\left(\frac{a}{b}-\lambda_i\right)-\left(\frac{a}{b}-\lambda_i\right)^2\right) \tag{7.50}$$

which look more complex than (7.18). This is due to the fixed endpoints.

Optimising (7.50) with respect to p_i one gets

$$p_i = \frac{4a}{9b}+\frac{3c_i}{9}+\frac{2\lambda_i}{9} \pm \frac{1}{18}\sqrt{36\left(\frac{a}{b}-c_i\right)^2 - 24\left(\frac{a}{b}-c_i\right)\left(\frac{a}{b}-\lambda_i\right)+34\left(\frac{a}{b}-\lambda_i\right)^2} \tag{7.51}$$

Note that, even if the λ_i in the right hand sides depend on prices, the dependence is crosswise, so (7.51) indeed provide explicit solutions for the mill prices. Again the second derivative has the sign of the root term of (7.51),

$$\frac{\partial^2 \Pi_i}{\partial p_i^2} = \pm 2\sqrt{36\left(\frac{a}{b} - c_i\right)^2 - 24\left(\frac{a}{b} - c_i\right)\left(\frac{a}{b} - \lambda_i\right) + 34\left(\frac{a}{b} - \lambda_i\right)^2} \quad (7.52)$$

so the root with the minus sign again corresponds to the maximum. It would now be possible to set up a dynamical system, using (7.37) and (7.38) and the smaller root of (7.51) with (7.48) and (7.49) substituted.

Two Identical Firms on a Fixed Interval

One may expect things to be simplified if the firms are assumed to be identical. In the model they are different only due to one single parameter, the marginal cost. Hotelling, like Cournot, as well as Palander, Wald, and von Stackelberg in some of their models, took marginal cost to be zero, One does not need to go that far in order to attain some simplification. Non-zero costs as well as elastic demand are things one would like to keep, so just put

$$c_1 = c_2 = c. \quad (7.53)$$

Cournot Equilibrium

Consider first the case of equilibrium. Then equations (7.39), (7.40), (7.48), (7.49), and (7.51), the last equation taken with the minus sign before the root term, hold as a simultaneous system.

Skipping messy derivations, just from the symmetry of this case, one can conjecture that

$$-x_1 = x_2 = x \quad (7.54)$$

$$p_1 = p_2 = p \quad (7.55)$$

and

$$\lambda_1 = \lambda_2 = \lambda \quad (7.56)$$

would solve the problem.

Indeed, substituting (7.54) and (7.55) in (7.37) and (7.38) reduces both equations (7.37) and (7.38) to the same, i.e., to

$$x = \frac{2}{3} - \frac{1}{3k}\left(\frac{a}{b} - p\right) \tag{7.57}$$

From (7.43) and (7.49), given (7.54) and (7.56),

$$\lambda = p + k(1 + x) \tag{7.58}$$

holds.

It is more messy to check out price. Substitution from (7.55) and (7.58) in (7.51) makes mill price (for both firms) come in the right hand side, even under the root sign, so it is necessary to rearrange, take squares to remove the root, and solve a new quadratic equation again choosing the correct root.

The most handy expression for price makes use of a compound parameter defined as the difference between maximum price and marginal cost divided by the transportation cost, i.e.,

$$\kappa = \frac{1}{k}\left(\frac{a}{b} - c\right) \tag{7.59}$$

Using this (7.51), selecting the proper sign, can be rearranged as.

$$\frac{1}{k}\left(\frac{a}{b} - p\right) = \frac{3}{5}\kappa + \frac{3}{10}\sqrt{34 + 4\kappa^2 - 8\kappa} - \frac{8}{5} \tag{7.60}$$

The quadratic form under the root sign is positive definite, so there is never any question of the root (7.60) not being real. Further, one would want p according to (7.60) to be larger than c, so that the firms obtain a positive profit. A numerical check shows that there is a wide interval

$$\kappa \in \left[5 - \tfrac{3}{2}\sqrt{10}, 5 + \tfrac{3}{2}\sqrt{10}\right] \tag{7.61}$$

such that this indeed is the case. Hence, the Cournot equilibrium for Hotelling's two identical firms on a fixed interval indeed has a reasonable solution when demand is elastic.

It may be interesting to check out two cases. First, when do the firms locate in the same point even according to the inner solution, i.e., when is $x = 0$? From (7.57),

$$x = 0 \quad \Rightarrow \quad \frac{1}{k}\left(\frac{a}{b} - p\right) = 2 \tag{7.62}$$

However, one would prefer have the condition expressed in terms of marginal cost in stead of in terms of price. To eliminate price, substitute from (7.60) to obtain

$$\kappa = 2.75 \tag{7.63}$$

Second, when do the firms locate in each in the centre of their market areas, i.e., when is $x = 0.5$? Again, from (7.57)

$$x = \frac{1}{2} \quad \Rightarrow \quad \frac{1}{k}\left(\frac{a}{b} - p\right) = \frac{1}{2} \tag{7.64}$$

and, using (7.60) to eliminate price,

$$\kappa = 0.75 \tag{7.65}$$

From Sect. 7.3 this is precisely the case of disjoint monopolies. In true duopoly, the firms will locate more close, as already Hotelling conjectured in his statement: With elastic demand "*the tendency to establish business excessively close will be less marked*" (than in the case of inelastic demand), (but the competitors) "*would not go as far ... as public welfare would require*" (i.e., locating at the quartiles of the interval) due to the "*tempting intermediate market*".

Both the extreme cases, clustering in the midpoint according to (7.62) and (7.63), and the socially optimal location at the quartiles according to (7.64) and (7.65) fall in the interval for positive profits as stated in (7.61).

Note that the case of clustering according to (7.62) and (7.63) is when the inner optima lead to this solution, which according to (7.59) happens at low transportation costs and/or low marginal production cost.

Even if the parameter values of the model do not lead to this situation, the possibility remains that any one of the competitors deliberately locates next to the other in an attempt to take the whole market. This is equivalent to Bertrand's price cutting argument and is, of course not covered by the analysis of inner optima above. The significance of this analysis was to show that such inner optima without crowding do exist for a wide variety of parameters, which they did not in Hotelling's original model with inelastic demand.

Dynamic Adjustment

The dynamic associated with this process is four dimensional, involving best reply adjustments of location and mill price of both competitors. To state this process, the relations from Sect. 7.5.1 will not do; it is necessary to get back to the more general formulas (7.37) and (7.38) for locations, (7.51) for prices, and (7.48) and (7.49) for the auxiliary variables.

The dynamic system, using period attributions for the variables, now reads

$$x_1(t+1) = \frac{x_2(t)-4}{5} + \frac{p_2(t)-3p_1(t)}{5k} + \frac{2a}{5kb} \tag{7.66}$$

$$x_2(t+1) = \frac{x_1(t)+4}{5} + \frac{3p_2(t)-p_1(t)}{5k} - \frac{2a}{5kb} \tag{7.67}$$

$$p_1(t+1) = \frac{4a}{9b} + \frac{3c}{9} + \frac{2\lambda_1(t)}{9} - \frac{1}{18}\sqrt{\Delta_1(t)} \tag{7.68}$$

$$p_2(t+1) = \frac{4a}{9b} + \frac{3c}{9} + \frac{2\lambda_2(t)}{9} - \frac{1}{18}\sqrt{\Delta_2(t)} \tag{7.69}$$

where, from (7.48) and (7.49),

$$\lambda_1(t) = p_2(t) + k(1 + x_2(t)) \tag{7.70}$$

$$\lambda_2(t) = p_1(t) + k(1 - x_1(t)) \tag{7.71}$$

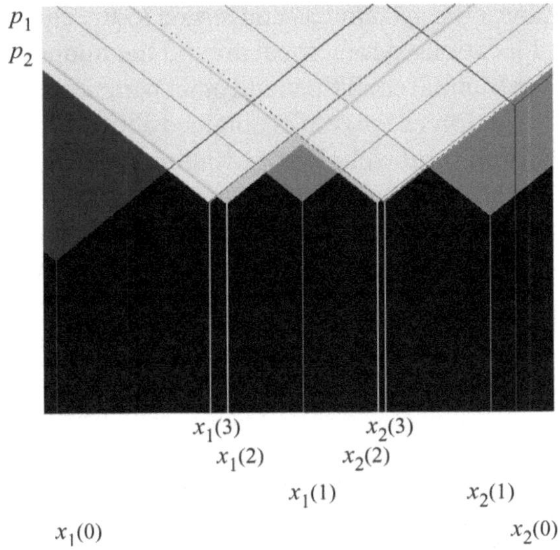

Fig. 7.3 Convergent dynamic process for location and mill price choice in the Hotelling case with elastic demand

and further, defined for convenience,

$$\Delta_1(t) = 36\left(\frac{a}{b} - c\right)^2 - 24\left(\frac{a}{b} - c\right)\left(\frac{a}{b} - \lambda_1(t)\right) + 34\left(\frac{a}{b} - \lambda_1(t)\right)^2 \quad (7.72)$$

$$\Delta_2(t) = 36\left(\frac{a}{b} - c\right)^2 - 24\left(\frac{a}{b} - c\right)\left(\frac{a}{b} - \lambda_2(t)\right) + 34\left(\frac{a}{b} - \lambda_2(t)\right)^2 \quad (7.73)$$

It was proved in Puu and Gardini (2002), using results due to Fairbrother (1973), that the system (7.66)–(7.69) is a contraction, i.e., not only is the Cournot equilibrium as defined in Sect. 7.5.1 locally stable, but the whole dynamic adjustment process leads to the Cournot equilibrium state. The proof is a bit lengthy, so it will not be repeated here.

In Fig. 7.3 the dynamic contraction process (7.66)–(7.73) is illustrated. The convergence is so fast that though 15 moves were run it is only possible to see the initial state and two intermediate ones before the system has converged to the equilibrium state. The parameters are $a = 2$, $b = 1$, $c = 0.5$, $k = 1$, and the initial conditions $x_1 = -0.95, x_2 = 0.9, p_1 = 0.75, p_2 = 1.5$. Hence quite different initial prices and locations close to the endpoints of the interval were chosen.

Yet, in only four moves the system has converged to the equilibrium state with equal prices and locations symmetrical around the midpoint. The same scenery is seen from other initial conditions and other parameter combinations.

This global stability reinforces the results obtained above. Not only does an inner Cournot equilibrium for the choice of location and mill price for both duopolists exist, but it is globally stable in the sense that every dynamic process from whatever initial conditions leads to the Cournot equilibrium.

Hence, quite as Hotelling himself indicated, his much celebrated paradox is removed once one introduces elastic demand.

References

d'Aspremont, C., Gabszewitsc, J. J., & Thisse, J. F. (1979). On Hotelling's "stabilty in competition". *Econometrica, 47*, 1145–1150.

Beckmann, M. (1968). *Location Theory*. New York: Random House Inc.

Fairbrother, R. W. (1973). Simplified Samuelson conditions for cubic and quartic equations. *The Manchester School of Economic and Social Studies, 41*, 396–400.

Hotelling, H. (1929). Stability in competition. *Economic Journal,* 39: 41–57.

Lerner, A. P., & Singer, H. W. (1937). Some notes on duopoly and spatial competition. *Journal of Political Economy, 45*, 145–186.

Puu, T., & Gardini L. (2002). Hotelling type duopoly and oligopoly. In T. Puu, I. Sushko, (Eds.), *Oligopoly Dynamics* (pp. 265–310). Berlin: Springer.

Smithies, A. (1941a). Optimum location in spatial competition. *Journal of Political Economy, 49*, 423–439.

Smithies, A. (1941b). Monopolistic price policy in a spatial market. *Econometrica, 9*, 63–73.

Index

T. Puu, *Oligopoly: Old Ends – New Means*, DOI 10.1007/978-3-642-15964-0,

FSC
www.fsc.org
MIX
Papier | Fördert
gute Waldnutzung
FSC® C083411

Kolibri 360 GmbH
Ferdinand-Jühlke-Straße 7
99095 Erfurt, Deutschland
produktsicherheit@kolibri360.de